J. A. Connelly

# Integrated Circuit Fabrication Technology

# Integrated Circuit Fabrication Technology

Andrew Veronis

Reston Publishing Company, Inc.
A Prentice-Hall Company
Reston, Virginia

© 1979 by Reston Publishing Company, Inc.
*A Prentice-Hall Company*
Reston, Virginia 22090

All rights reserved. No part of this book may be reproduced in any way, or by any means, without permission in writing from the publisher.

**Library of Congress Cataloging in Publication Data**

Veronis, Andrew.
  Integrated circuit fabrication technology.

  Includes index.
  1. Integrated circuits.   I. Title.
TK7874.V47          621.381'73          78–31381
ISBN 0–8359–3092–0

10 9 8 7 6 5 4 3 2 1

Printed in the United States of America

**This book is dedicated to my wife**

# Contents

**Preface, xi**

**chapter one    Introduction to Integrated Circuits, 1**

    1.1   Types of Integrated Circuits, 1
    1.2   Silicon Planar Process, 1
    1.3   Integrated-Circuit Components, 3
    1.4   Monolithic Circuits, 4
    1.5   Circuit Element Design, 7
    1.6   Practical Considerations, 8
    1.7   More Complex Circuits, 9
           Exercises, 12

**chapter two    Fabrication Materials, 13**

    2.1   Silicon, 13
    2.2   Silicon Dioxide, 16
    2.3   Sapphire, 17
    2.4   Gold and Aluminum, 18
    2.5   Epoxies, 18
    2.6   Ceramic and Glass, 19
    2.7   Kovar, 19
    2.8   Packing Materials, 20
           Exercises, 21

## chapter three     Integrated Resistors, 23

- 3.1 Silicon Resistors, 23
- 3.2 Diffused Resistors, 24
- 3.3 Epitaxial Layer Resistors, 29
- 3.4 Vapor-Deposited Resistor, 29
- 3.5 Thin-Film Resistors, 30
- 3.6 Practical Design, 30
- Exercises, 33

## chapter four     Integrated Capacitors, 34

- 4.1 Introduction, 34
- 4.2 Diffused Capacitors, 34
- 4.3 Epitaxial Capacitors, 35
- 4.4 Design of Diffused Capacitors, 36
- 4.5 MOS Capacitors, 39
- 4.6 Design of MOS Capacitors, 40
- Exercises, 41

## chapter five     Integrated Diodes and Transistors, 42

- 5.1 Basic Diode Concepts, 42
- 5.2 Diode Applications, 44
- 5.3 Bipolar Transistors, 46
- 5.4 Parasitic Elements, 48
- 5.5 Integrated-Transistor Structures, 51
- 5.6 Design Considerations, 53
- 5.7 Integrated-Transistor Types, 54
- 5.8 Design Examples, 59
- Exercises, 70

## chapter six     Metallic-Oxide-Semiconductor Circuits, 71

- 6.1 General, 71
- 6.2 Basic MOS Transistor, 71
- 6.3 MOS Applications, 73
- Exercises, 76

## chapter seven     Basic Design Rules and Equations, 77

- 7.1 Introduction, 77
- 7.2 Mathematical Equations, 77

- 7.3 Practical Designs, 81
- 7.4 Practical Design of Resistors, 82
- 7.5 Practical Design of Integrated Capacitors, 84
- 7.6 Practical Design of Transistors, 86
  Exercises, 94

## chapter eight   Photoengraving and Mask Fabrication, 95

- 8.1 Introduction, 95
- 8.2 Artwork, 95
- 8.3 Mask Design, 96
- 8.4 Photoreduction, 97
- 8.5 Masks, 97
  Exercises, 102

## chapter nine   Linear Integrated Circuit Design, 103

- 9.1 Introduction, 103
- 9.2 Capacitor Applications, 104
- 9.3 Lateral PNP's, 104
- 9.4 Super-beta Transistors, 108
- 9.5 Schottky-Clamped Structures, 109
- 9.6 Pinch Resistors, 109
- 9.7 Differential Amplifier, 112
- 9.8 Operational Amplifier Input Stage, 113
- 9.9 Additional Stages, 114
- 9.10 Biasing Circuits, 117
- 9.11 Direct-Current Bias Currents, 120
  Exercises, 125

## chapter ten   Digital Circuit Design, 126

- 10.1 Introduction, 126
- 10.2 Important Terminology, 126
- 10.3 Transistor-Transistor Logic (TTL), 131
- 10.4 Transistor-Transistor-Logic Outputs, 133
- 10.5 Input and Output Impedance, 135
- 10.6 Input and Output Voltages, 136
- 10.7 Input and Output Currents, 137
- 10.8 Low Power and Schottky TTL, 137
- 10.9 Emitter-Coupled Logic (ECL), 139

## Contents

- 10.10 Additional ECL Configurations, 141
- 10.11 Integrated-Injection Logic ($I^2L$), 143
- 10.12 Advanced Logic Circuits, 143
  - Exercises, 147

### chapter eleven  MOS and CMOS Circuit Design, 148

- 11.1 Introduction, 148
- 11.2 Basic Operation of the MOS Transistor, 148
- 11.3 Interpretation of Symbols, 151
- 11.4 Basic MOS Circuits Common Source, 152
- 11.5 Integrated Circuits, 153
- 11.6 MOS Memory Designs, 158
  - Exercises, 161

### chapter twelve  Large-Scale Integration, 162

- 12.1 Introduction, 162
- 12.2 Integrated-Injection Logic ($I^2L$), 163
- 12.3 MOS/CMOS LSI, 165
- 12.4 The (111) Standard Process (P-Channel Enchancement Mode), 167
- 12.5 The (100) Process, 170
- 12.6 Ion Implantation, 170
- 12.7 Memory Circuits, 170
  - Exercises, 175

### chapter thirteen  Integrated-Circuit Applications, 176

- 13.1 The Operational Amplifier, 176
  - Exercises, 187

**Index, 188**

# Preface

Never has the electronic industry seen such a fast and extensive development as with the birth of the integrated circuit. Complex circuits which required large space are now a routine reality. Thanks to the integrated circuit, medical equipment, space electronics, and all phases of everyday life have improved considerably.

The task that today's engineer has to face is twofold. He not only must know how to design the electronic circuit in integrated form, but he must also have a fairly comprehensive knowledge of how this circuit will be fabricated and tested. This book provides substantial knowledge in the field of integrated circuit fabrication.

I would like to express my sincere thanks to the major semiconductor manufacturers such as Texas Instruments, National Semiconductors, Motorola, and others, for allowing me to use their valuable knowledge in this book. I also wish to thank everyone who helped during the production of the text.

# chapter one
# Introduction to Integrated Circuits

## 1.1 Types of Integrated Circuits

Several types of integrated circuits exist; they are identified according to their process technique. Film circuits and semiconductor monolithic circuits are two types of circuits used at present.

In the film technique, one or more elements or compounds are deposited selectively on a passive substrate (i.e., glass or ceramic) to form resistors, capacitors, and electrical interconnections. Resistors are obtained by depositing a material, such as nichrome, tantalum, or cermet, that has a relatively high sheet resistance, and by forming rectangular areas of the film, employing suitable masking techniques, which are discussed later. Similarly, capacitors are obtained by selective deposition of a suitable dielectric, such as silicon monoxide or aluminum silicate, over a metallic film, and then subsequently depositing a second metallic film over the dielectric within a prescribed area. Anodized tantalum films are also used to obtain capacitors.

## 1.2 Silicon Planar Process

The silicon planar process will be discussed in detail later, but a brief description follows. The basis of the silicon monolithic integrated circuit is the silicon-double-diffused planar-type transistor. Two basic process steps involved in fabricating such a structure are (1) gaseous diffusion of impurity atoms into a semiconductor wafer, and (2) photolithographic techniques that permit selective diffusion of impurities into only certain portions of the wafer. A layer of silicon dioxide on the surface of a silicon wafer effectively blocks diffusion of impurity atoms to the silicon. Thus, if the silicon wafer is partly exposed to the diffusant, and partly covered with silicon dioxide, diffusion will occur only in the unoxidized portion of the wafer. In practice, a uniform layer of silicon dioxide is grown over the wafer; then photolithographic techniques are used to remove the oxide from portions of the wafer where diffusion is desired.

Thus, to make a planar-type double-diffused *npn* transistor, *p*-type impurity atoms are diffused into areas of the wafer where base regions are desired, and then *n*-type impurities are diffused into the base regions to form emitter regions. The number of base regions that can be formed on a single wafer depends upon the size of the transistor and its application (i.e., high-frequency or high-power), but normally several hundred to several thousand transistors are made on one wafer. *PNP* transistors can be made in the same manner, by using a *p*-type wafer and *n*- and *p*-type base and emitter diffusants, respectively.

After transistor structures have been fabricated by this process, contacts are made to the emitter, base, and collector regions of each structure by evaporating a metal film over the surface of the transistor, again using photolithographic techniques to define the areas of contact.

From an electrical circuit point of view, the salient feature of this fabrication process is that a number of *pn* junctions can be described by the fundamental junction equation, relating junction voltage $V$ to current $I$ through the junction:

$$I = I_s \left(\exp \frac{8 \text{ eV}}{kT} - 1\right)$$

where $I_s$ is the saturation current of the junction and $\left(\frac{kT}{qe}\right)$ is the thermal voltage (25 millivolts, mV, at room temperature).

When two *pn* junctions having a common *p* or *n* region are located close together, carriers from one junction can interact with carriers of the adjacent junction, so that the three-electrode structure, *pnp* or *npn*, has potential gain. A well-known equivalent electrical circuit for such a *pnp–npn* structure is shown in Fig. 1.1. Here, $\alpha_N$

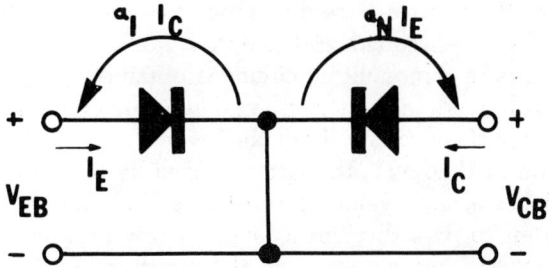

Figure 1.1  Diode equivalent circuit for junction transistor.

and $\alpha_I$ denote the normal and inverse current transfer ratios (each of which is normally less than 1).

## 1.3 Integrated-Circuit Components

The process for the fabrication of transistors is described above. Diodes in an IC can be fabricated by employing one of several combinations of emitter–base or collector–base junctions of a transistor structure. For example, the collector–base junction can be used with the emitter open-circuited, or the emitter–base diode can be used with the collector open-circuited, as shown in Fig. 1.2.

It follows that there are five such possible combinations, leading to five types of diodes, each having its particular electrical charac-

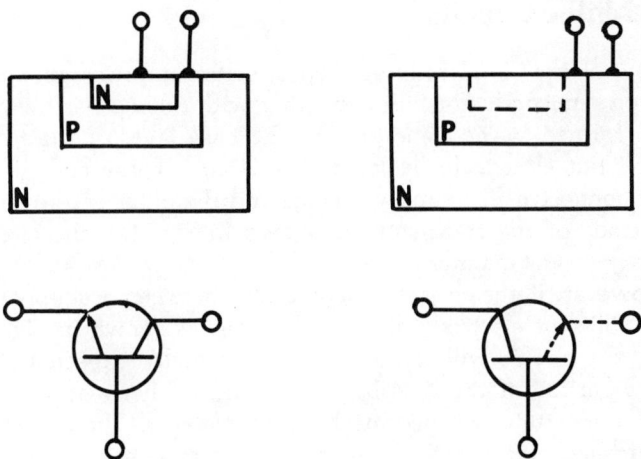

Figure 1.2  Diode configuration.

teristics (e.g., the emitter–base diode may be used with the collector–base terminal short-circuited, and so on).

Resistance in a monolithic circuit is obtained by the use of one of the diffused layers in the structure. The value of the resistor is determined by the sheet resistivity of the layer multiplied by the length-to-width ratio of the layer; the latter is fixed by proper design of the masking. Hence a large value of resistance is obtained by constraining the current to pass through a long, narrow portion of a diffused layer. Any of the three layers of the structure (corresponding to emitter, base, or collector regions of the transistors) can be used for a diffused resistor. The most commonly used layer is the base region, in which the resistivity generally is of the order of 100–200 ohms ($\Omega$)/square. Thus, to obtain a 1-kilohm (k$\Omega$) resistor, in this case, the resistor must be 10 times as long as it is wide.

Capacitance can be obtained by using the transition capacitance of a reverse-biased *pn* junction. The value of the capacitor is directly proportional to the area of the *pn* junction, which again is determined by the proper design of the masking. Alternatively, in a slight variation of the monolithic process, capacitance can be obtained by the use of the silicon dioxide layer, which normally is employed for the proper diffusion technique. Here the oxide acts as a dielectric sandwiched between the semiconductor substrate and a deposited metallic layer of suitable area. The metallic layer is that normally employed in the fabrication process for interconnections between circuit elements.

## 1.4 Monolithic Circuits

One of the simplest integrated circuits comprises an *npn* transistor in a semiconductor block together with a long, narrow diffused *p* layer (formed at the same time as the base layer of the transistor structure, but electrically isolated therefrom). If one end of the resistor is connected to a supply voltage and the other end to the emitter electrode of the transistor, as shown in Fig. 1.3, the circuit behaves in the usual manner.

However, if the resistor is connected between a supply voltage and the collector electrode of the transistor, as shown in Fig. 1.4, in general, the circuit will *not* behave as normally expected. In particular, a portion of the distributed *pn* junction between the resistor and the *n* substrate will become forward biased as the collector current increases, and effectively bypass the series resistor. This illustrates the point that, in a monolithic integrated circuit, there is no such thing as a *pure* circuit element. There is always a distributed diode effect between one diffused layer and its neighbors. The dis-

Monolithic Circuits 5

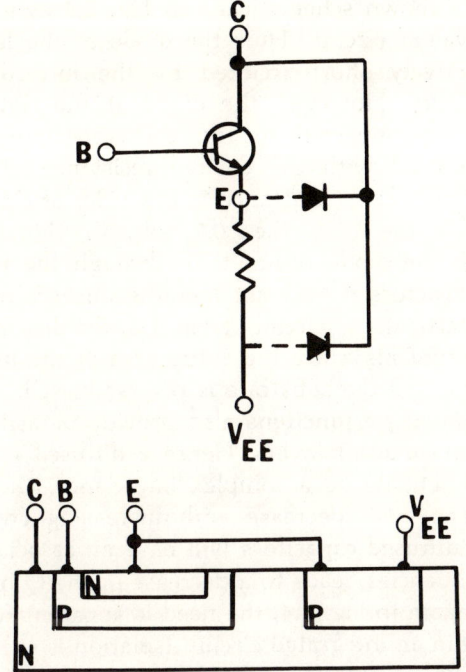

Figure 1.3 Normal operation of the circuit.

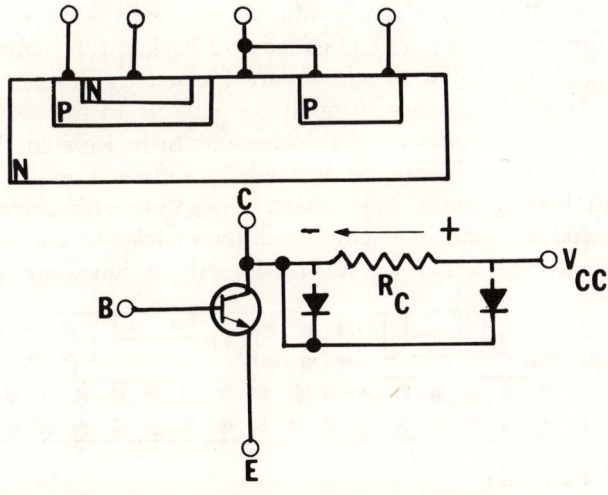

Figure 1.4 Abnormal operation of the circuit.

tributed effect is shown schematically in Fig. 1.4 as a three-element pi-lumped equivalent circuit. Thus the diode at the left end of the resistor is effectively short-circuited by the interconnection, and creates no difficulty. However, the diode at the right end of the resistor, which is connected to the collector-supply terminal, is effectively in parallel with the desired resistance. As the voltage builds up across the series resistance to the order of the forward conduction voltage of the diode (i.e., 0.7 volt, V), the current will be shunted through the diode rather than through the resistance, and the voltage drop across $R$ will not increase linearly with increasing current. In the particular $C$ circuit of Fig. 1.3, the distributed $pn$ junctions cause no problems, since the entire area of the $pn$ junction between the resistor and the substrate is reverse biased.

The distributed $pn$ junctions also provide parasitic capacitance coupling between adjacent layers. Hence a diffused resistor at high frequencies will actually be a complex impedance, and the effective value of resistance will decrease with increasing frequency. In a similar fashion, diffused capacitors will have an associated resistance that, at high frequencies, leads to a decrease in the $Q$ of the element.

Referring again to Fig. 1.4, the need is apparent for isolation of components within an integrated circuit. Isolation is even more necessary in more complex circuits, such as circuits containing more transistors whose collectors must be at different potentials.

To obtain the required isolation in a monolithic fabrication, several different techniques have been developed. Historically, the first, and still most widely used, technique is that of $pn$ junction isolation. In this case, a four-layer semiconductor structure is employed. For example, $npn$ transistors are embedded in a basic $p$-type substrate, as shown in Fig. 1.5. Isolation *islands* are formed by diffusing $p$-type impurities from the surface through the $n$ layer to the $p$-type substrate to form an island around the elements to be isolated. Then the second $p$ and $n$ layers can be formed by diffusion in the normal fashion. By biasing the $p$-type substrate negative with respect to all other electrodes, circuit elements in different islands are effectively isolated from each other by reverse-biased $pn$ junctions. Parasitic

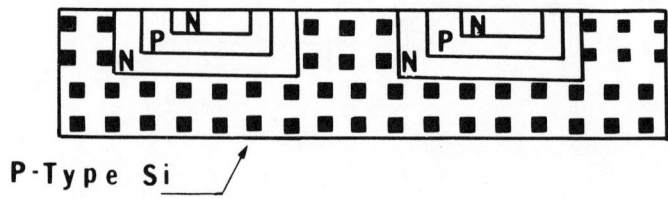

Figure 1.5 Integrated pn-junction isolation.

coupling in the form of junction capacitance still exists, however, and this leads to some decrease in high-frequency, or high-speed, performance of circuits employing this type of isolation.

## 1.5 Circuit Element Design

Although, in principle, a number of different layers could be diffused into separate portions of the circuit to provide desired passive-component values, in practice, normally one series of diffusions is used. Thus the resistivities and depths of the layers are determined to a large degree by design considerations for the transistor. Consequently, the values of the passive components must be derived from the fixed properties of these layers by suitable control of the geometry during the masking processes.

Therefore, circuit design of monolithic circuits is principally an exercise in map making, or topography, once the basic theoretical design and breadboarding approach have been determined. The size of the components in a circuit is determined on the one hand by the tolerances obtainable in the manufacturing processes, and on the other hand by the current-handling requirements of the structure. High-frequency requirements generally dictate small capacitances, which, in turn, require minimum areas and maximum layout efficiency. The smallest dimension obtainable in a producible circuit has been steadily decreasing over the years as a result of improvements in photolithographic techniques. For example, as shown in Fig. 1.6, the size of a transistor in a recent circuit is about one tenth the size of the smallest transistor in the late 1950s.

These same improvements are also reflected in the size of passive components. For example, the width of a diffused resistor is at present smaller than, say, in 1958, so that for a fixed length-to-

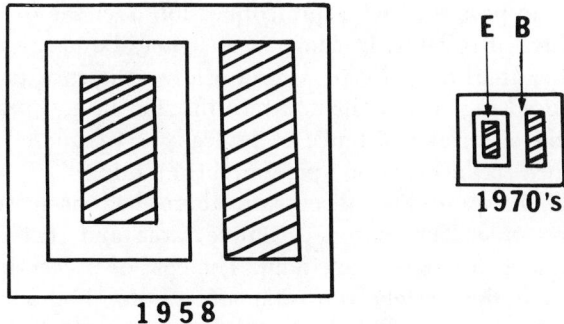

Figure 1.6 Dimensional tolerances in ICs.

width ratio by the value of resistance required, less area is required now.

Unfortunately, there are no straightforward, rigorous procedures for designing a monolithic integrated circuit. Given a specific circuit, there may be several equally valid designs. However, in laying out a circuit, certain *ground rules* must be followed, and many of these are functions of the state of the art, which is constantly changing. For example, the minimum allowable tolerances in fabricating a monolithic structure set a lower limit to the emitter width of the transistor. The emitter length depends upon the current-handling requirement of the transistor. The overall size of the transistor is then set by the emitter dimensions, together with the effect of fabrication tolerances in determining collector–base junction and collector–substrate junction cross-sectional areas.

## 1.6 Practical Considerations

In addition to the area occupied by the circuit elements and the isolation islands, space also must be allotted for areas between isolation islands, and for the placing of bonding areas, or pads, where connections are made between the pins of the package and the corresponding metallized electrodes of the silicon circuit. The number of the bonding pads, naturally, depends upon the number of connections required between the silicon circuit and the outside world, which in turn is a function of the complexity and application of the circuit.

Additional areas might be required in some cases for metallized intraconnections in the surface of the silicon wafer between circuit elements or between a circuit element and a bonding pad. Normally, the intraconnections pass over edges or areas between isolation islands. Circuit crossovers should be minimized as much as possible in the circuit layout. The most convenient method of providing a crossover is to pass the metal intraconnection over the oxide layer on top of a diffused resistor. In some cases it may be necessary to cross under, rather than over, by providing a low-resistance path through an $n+$ diffused region (by the emitter diffusion in an *npn* structure). This provides a series resistance of typically 3 $\Omega$/square and a parasitic capitance on the order of 1 picofarad (pF)/mil$^2$.

In all the preceding discussion there has been considerable emphasis on *area*. The reason is simple; area and cost are directly related in a silicon monolithic circuit. The cost of processing a silicon wafer through the various diffusion and photolithographic steps is more or less independent of the type of circuit or devices in the silicon, assuming that the circuit is fabricated under normal manufacturing tolerances. Hence the larger the wafer used for a given circuit, the more expensive is the circuit. At the same time, of course, the larger the circuit, the more elements that can be incorporated into the circuit, that is, the more complex the circuit.

There are practical limits to circuit size in both directions. In the limit of small size, the silicon monolithic circuit degenerates into individual circuit elements that require a large number of separate packages and separate terminals, thus losing the advantage of low cost and high reliability of integrated circuits. At the other extreme, with very large circuits, there is a problem known as *yield*. In any silicon wafer that has been processed to form an integrated circuit, there are defects, normally randomly distributed over the wafer. The larger the individual circuit, the more likelihood that any one defect will fall within a given circuit and, hence, render it defective and unsuitable for sale. For example, if there were 20 defects randomly scattered over a wafer containing 200 circuits, the yield at this stage of the process would be 90 percent (i.e., 90 potentially good circuits available out of the 100). However, if the circuits were four times as large, so that there were only 50 per wafer, the potential yield would be only 40 percent, and the larger circuit effectively would cost two and one quarter times as much as the smaller circuit at this stage of the process. Between the two limits, there is an optimum circuit size.

## 1.7 More Complex Circuits

The foregoing discussion has been limited to the simplest types of circuits containing (by implication) *npn* transistors, diffused resistors, and oxide capacitors. More sophisticated circuits can be made by simple extensions of the basic techniques described.

One modification employed in many circuits is the use of an additional, heavily doped, $n+$ region beneath transistor collector regions, as shown in Fig. 1.7. In normal discrete diffused planar-type tran-

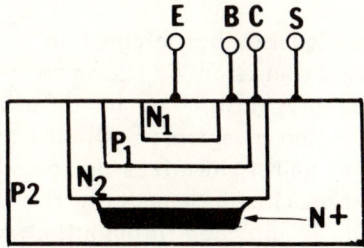

Figure 1.7  Isolated transistor with reduced collector series resistance.

sistors, the collector contact is made to the bottom of the wafer, and emitter circuit flows directly through the emitter–base collector structure. On the other hand, in monolithic circuits, transistor collector current must flow transversely through the relatively thin collector region en route to the collector contact on top of the overall wafer. This leads to an increased resistance in series with the collectors and, hence, increases the collector–emitter saturation voltage of the transistor. The use of an additional $n+$ layer beneath the collector provides a much lower resistance path, and decreases the saturation voltage. This, of course, is obtained at the expense of an additional process step, which, in turn, may decrease yield and somewhat increase cost. However, it is inescapable using epitaxial processing techniques featuring layers on the order of 5–10 micrometers ($\mu$m) in depth.

In the fabrication process just described, a single-polarity type of transistor (e.g., *npn*) normally is used. In some types of circuits, complementary transistors (both *npn* and *pnp*) are desirable. To obtain both types of transistors in a monolithic circuit, several approaches can be used. Actually, *pnp* transistors already exist in all monolithic circuits containing *npn* transistors and using junction isolation, as described previously. There is one obvious disadvantage in this type of *pnp* transistor, however, in that one *p* region, normally the collector, is electrically connected to the substrate or ground. If the circuit design is such that the collector of the *pnp* transistor is, indeed, connected to the ground, as in an emitter follower, for example, then this type of structure would be satisfactory. To eliminate this disadvantage, a five-region structure may be used to obtain both *pnp* and *npn* transistors that are electrically isolated (through *pn* junctions) from ground. This, of course, requires still another process step relative to the basic monolithic circuit. It should be pointed out that a compromise between optimizing the *npn* and *pnp* elements is inevitable, since it is impossible to carry out the doping and diffusion steps independently.

*PNP* transistors also can be obtained in a monolithic circuit normally containing *npn* transistors by taking advantage of the lateral transistor effect noted previously (i.e., by diffusing two *p* regions close together in a common *n* region). Such structures normally also have fairly low current gains, since it is not possible to control the spacing between adjacent circuit elements to the same small dimensions as can be done in the direction through the wafer. Nevertheless, by additional circuit "tricks," it is possible to provide complementary-type circuits. For example, the *pnp* transistor having low current gain can be connected directly to an *npn* transistor to provide useful current gain.

A third modification of the conventional monolithic circuit structure to improve circuit performance (at the expense of additional process steps) is to combine thin-film techniques with the silicon monolithic structure. The thin-film process is described later. It turns out that thin-film-deposited resistors have better electrical characteristics than diffused resistors. For example, their temperature variation is approximately 10 times less than that of a diffused resistor, and their Q at high frequencies is somewhat better. In addition, by proper control of the film material it is possible to obtain higher values of resistance per unit area.

This chapter provides a brief but essential survey of the vast field of integrated circuits. Most of the topics touched on here are discussed in detail in the chapters that follow.

## Exercises

**1.1** Provide a series of guidelines for the designing of monolithic circuits.

**1.2** Describe how a capacitor can be obtained in a monolithic IC.

**1.3** Applying the junction equation, find $I$ when $I_s$ is 500 milliamperes (mA).

**1.4** How is a double-diffused *npn* transistor fabricated?

**1.5** Describe the same process for a *pnp* transistor, and outline any differences.

**1.6** Give the total resistance of a diffused resistor when it is 80 mils long and 2 mils wide.

**1.7** Provide a circuit for a *pnp–np* structure.

**1.8** Using the symbol of a transistor, provide the schematic for a diode as it would be shown in an integrated circuit form.

**1.9** Describe isolation and isolation islands.

**1.10** How is the main circuit connected to the package pins?

## chapter two
# Fabrication Materials

## 2.1 Silicon

Silicon is employed widely in the fabrication of integrated circuits, and, although we do not intend to get involved heavily in the physical and electrical characteristics of semiconductor materials, the reader should become familiar with certain properties of silicon that affect its performance in IC design and fabrication.

Although silicon is employed more widely in IC fabrication than germanium, silicon presents certain disadvantages. It has a much higher melting point, and, when molten, acts as a universal solvent. Its distribution coefficients are extremely unfavorable for certain impurities. On the other hand, some of these disadvantages that render silicon more difficult to apply in the fabrication of IC's also make it suitable to better performance. For example, due to the higher melting point, silicon can be fabricated into IC's capable of operation at higher temperatures.

One disadvantage with both silicon and germanium is that they are indirect gap semiconductors. Therefore, many electrooptical applications cannot be accomplished with monolithic silicon IC's.

As a group IV element, the crystal lattice is tetrahedral, resem-

## 14    Fabrication Materials

bling that of a diamond. The electrical properties of silicon depend on the crystal structure, the type, and the number of impurities it contains.

A rather popular method of producing single crystals for use in semiconductor devices is the Czochralski method or growing technique. This technique consists of withdrawing, or pulling, a crystal from a molten mass of material under controlled conditions, as shown in Fig. 2.1. There are a number of reasons for the popularity of this method, the most significant being the following:

1. The need for large-diameter crystals so that more devices can be obtained per slice with a minimum of handling.

2. The need for material heavily doped with highly doped volatile impurities. This is no longer true in silicon requirements for today's microcircuit technology, but in the early days of epitaxial growth and planar device design the need was more extensive.

3. The need for low-dislocation density substrates for epitaxial growth.

4. The economic factor affecting cost of crystals.

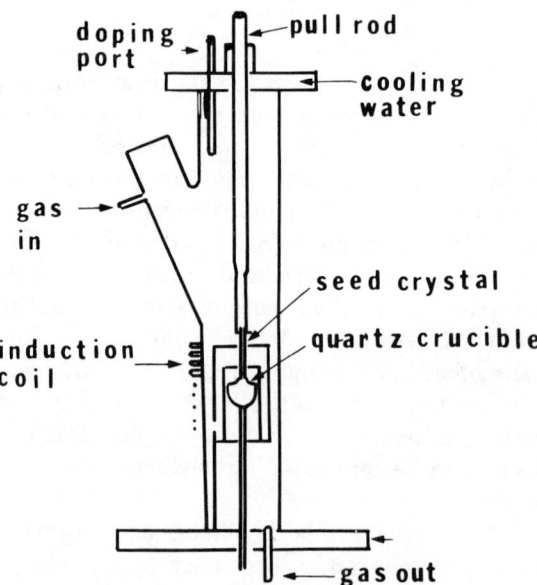

Figure 2.1   Czochralski furnace.

## Silicon

Approximately 75 percent of the single-crystal silicon used today is produced by the Czochralski process of crystal pulling; 25 percent is produced by the float zone refining technique.

Resistivity is the most important parameter of a semiconductor material, because it is a measure of the amount of doping impurity that has been added to the pure semiconductor material. The amount of doping impurity added determines the number of electrical-charge carriers. This, therefore, determines the current and voltage characteristics of the finished device.

*Resistivity*, or specific resistance, is the ratio of the voltage drop per unit length to the current per unit cross section, measured on a unit cube of the material. The relationship of resistivity to the properties of useful silicon material falling in the practical range for a device is:

$$\rho_p = \frac{1}{pq}$$

$$\rho_n = \frac{1}{nq}$$

where

$$p = N_A - N_D$$
$$n = N_D - N_A$$

$\rho_p, \rho_n$ = resistivity of $p$ and $n$ material, respectively
$\mu_p, \mu_n$ = drift mobility of holes and electrons, respectively
$N_A, N_D$ = density of acceptor and donor atoms, respectively
$q$ = electron charge

Upon studying these equations, it is concluded that (1) the resistivity of silicon is a function of the net electron or hole concentration, and (2) the only temperature-dependent qualities are mobilities $\mu_p$ and $\mu_n$.

Flat surfaces, such as sawed crystal ends or thin slices, are measured for resistivity by the four-probe method. Four equally spaced probes in an in-line array are placed in contact with the surface to be measured, as shown in Fig. 2.2. When the four-probe method is employed to measure the ends of large-diameter crystals, the volume is considered semiinfinite, and the resistivity is calculated as follows:

$$\rho = \frac{V}{I} 2\pi S$$

where

$V$ = voltage drop between inner probes
$I$ = current through outer probes
$S$ = uniform probe spacing

**16    Fabrication Materials**

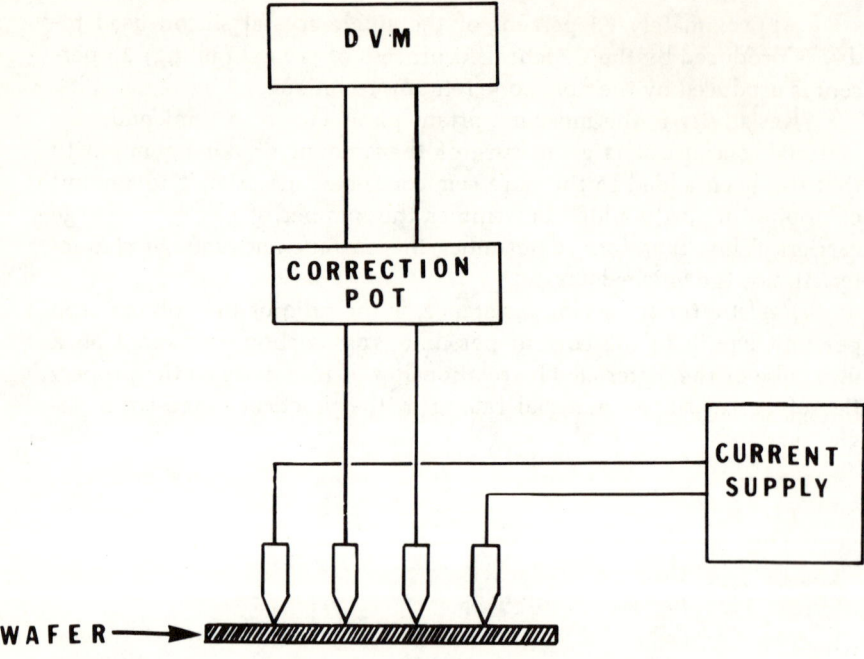

Figure 2.2    Four-probe resistivity measuring method.

Measurements at the end of crystals, at the center, and at one-half the radius from the center, establish the radial resistivity variation of the crystal. The term *radial gradient*, as used in the semiconductor industry, is the percentage of variation from the center to $R/2$.

The four-probe method has been found to provide best results if the probes are placed perpendicular to a diameter of the crystal ends or slice. In this position, the probes are perpendicular to the greatest change in resistivity in crystals rotated during growth.

## 2.2  Silicon Dioxide

The properties of silicon dioxide make it one of the most useful IC materials, mainly owing to the following reasons:

1. Continuous, uniform, closely adhering coating over crystalline silicon, and ease of controlling thickness. The thermal expansion coefficient is similar to that of silicon.

2. Ability to mask out the most important diffusants (i.e., phosphorus and boron).

3. Insulating material with useful dielectric.
4. Ease of etching or local removal.
5. Surface passivation or protection by tying up surface bonds of the silicon material. The junctions actually are formed under the oxide.

Silicon dioxide ($SiO_2$) is an effective barrier for many diffusants. For effective masking, the thickness ranges between 2000 and 10,000 angstroms (Å). The material has a shiny or glassy metallic appearance, and, depending on its thickness, appears pink, green, gray, or blue. The effectiveness of the oxide as a mask depends on both its thickness and the diffusion rate of the impurity. Glass is a good mask for phosphorus and boron, but is almost transparent for such diffusants as gallium. While preventing diffusion through its interface, $SiO_2$ permits diffusion into the silicon under it from an exposed surface. This has the effect of establishing the actual junction under $SiO_2$, thus providing a measure of protection or surface passivation where it is most required.

Hydrofluoric acid is used to etch $SiO_2$ from the regions where diffusion is to take place. A photosensitive substance, called *photoresist*, is applied; when exposed to ultraviolet light and properly developed, it will resist etching by the acid. The remaining unexposed photoresist may be washed off by a solvent after development.

## 2.3 Sapphire

It has been proved that complementary metallic-oxide semiconductor (CMOS) circuits, fabricated on sapphire substrates, present several advantages over those on silicon. They are faster, they have tighter, smaller circuit features, and, once the starting material has been prepared and polished, they are easier to build.

Sapphire shows a reduction in the junction and other parasitic capitance that occurs in silicon and slows down the performance of a circuit. Therefore, sapphire circuits are faster in operation. Sapphire circuits are smaller, about half the size of their silicon counterparts, because three levels of interconnect allow tight packing of the devices, and because the guardbands generally used for transistor isolation in silicon substrates are eliminated. The simplicity of the process makes the cost of CMOS-on-sapphire circuits competitive with, and perhaps lower than, that of bulk CMOS circuits, even though unpolished sapphire substrates in large quantities presently cost about six to seven times more than the others. Furthermore, a lower cost produces higher yields because of simpler processing and the fact that silicon-on-sapphire circuit operation is little affected by mask and oxide defects over inactive sapphire regions.

## 2.4 Gold and Aluminum

Gold and aluminum are the basic conductor materials in IC fabrication. The following are some of the advantages of gold:

1. It has a fairly high solid solubility, and thus the amount of gold introduced into an IC can be controlled over a wide range of values.
2. It does not form any compounds with silicon, and thus its behavior is free from anomalous effects that may be caused by the formation or decomposition of these compounds.
3. Although gold diffuses predominantly by an interstitial mechanism, over 90 percent of it terminates in substitutional sites and is electronically active.

Gold, combined with silicon to form a eutectic compositon, is widely used for soldering. Fine gold wire is used for wire bonding (bonding the points of the circuit to the pads, which are connected to the package terminals or pins).

Aluminum has found extensive use in IC's because of its ability to adhere both to silicon and silicon oxide. Functions performed by aluminum are the following:

1. Contact between exterior leads and individual function areas.
2. Use as the exterior leads (pins), mechanical supports, and thermal dissipators.
3. Resistance.
4. Use as capacitor plates.

In the latter, the metal, such as tantalum, is deposited to form one electrode.

## 2.5 Epoxies

Epoxies have found wide acceptance for the attachment of active semiconductors, chip capacitors, and resistors and other small parts of hybrid circuits. Their use provides the designer with a high degree of freedom when building complex circuits, and they allow employment of easily automated processes that provide good yields.

Of the several attachment methods currently in use (gold–silicon eutectic binding, soft solder, and conductive epoxy), epoxy is the fastest-growing technique. Its advantages are the following:

1. Conductor-pattern requirements are not demanding, and a wide range of conductor materials can be used. Epoxies allow use of low-cost palladium–silver or platinum–silver conductor patterns.

2. Epoxies can be applied by accurate pneumatic dispensing or screen printing, both low-cost, easily automated processes.

3. Low-temperature curing, as low as 150°C, removes the possibility of damage to delicate components, especially active devices. High yields of over 90 percent are readily attained.

4. Any wire-bonding method may be used with epoxy-attached devices. The epoxy can withstand thermocompression temperatures of even 320°C for short periods.

## 2.6 Ceramic and Glass

An insulating material with good conductivity and the ability to withstand high processing temperatures is desirable for IC fabrication. The ceramic elements alumina ($Al_2O_3$) and beryllia (BeO) are useful substrate materials. These same materials can be used as substrates for thin-film work after proper preparation or in packages for completed IC's. Alumina has approximately the same thermal conductivity as Kovar, described later.

Glass plates have been used as substrate material for many thin-film circuit fabrications. Tight flatness and uniformity specifications are satisfied at surprisingly low cost. Some thin-film circuit suppliers use large sheets of glass, which are cut into small individual circuits as a final step. Special glass mixtures with similar expansion coefficients as Kovar are marketed widely for glass-to-metal seal applications. Other special glass and plastic materials have been used to coat, protect, and seal completed IC's.

## 2.7 Kovar

Kovar is a nickel–iron alloy frequently used in glass-to-metal seal work. The Kovar is oxidized in making the glass-to-metal seal.

## 20    Fabrication Materials

Since the metal is easily oxidized, the external leads on an IC package are plated for ease in soldering and welding. Kovar is used in this application because it has the proper temperature coefficient of expansion to prevent the glass-to-metal seal from cracking during thermal cycling.

## 2.8  Packing Materials

Packaging materials are chosen for their desirable inert properties and their compatibility with silicon-processing temperatures. Matched coefficients of thermal expansion are also desirable.

Ceramic materials, such as alumina, have relatively good thermal conductivity as compared to glass, and a reasonable temperature coefficient to thermal expansion as compared to silicon. Glass has the advantage of being transparent, and in-process inspection can be performed on partially completed units. This can be a mixed blessing because of silicon's photosensitivity.

# Exercises

**2.1** Briefly explain some of the advantages of sapphire over silicon.

**2.2** Describe the use of a photoresist.

**2.3** Convert 1 Ω-cm to ohm-mils. Ohm-centimeters measure the bulk resistance of a semiconductor material (a 1-cm cube of 1 Ω between opposite faces). The problem is to determine the resistance between the opposite faces of a cube 0.001 in. (1 mil) on each edge, or to convert between English and metric system measurements.

**2.4** Using the value obtained in Exercise 2.3, what is the resistance in series with the collector of a conventional bottom contact transistor with an effective collector area of 2 by 3 mils and a wafer 5 mils thick under the collector? Use the accompanying illustration, and neglect spreading effect.

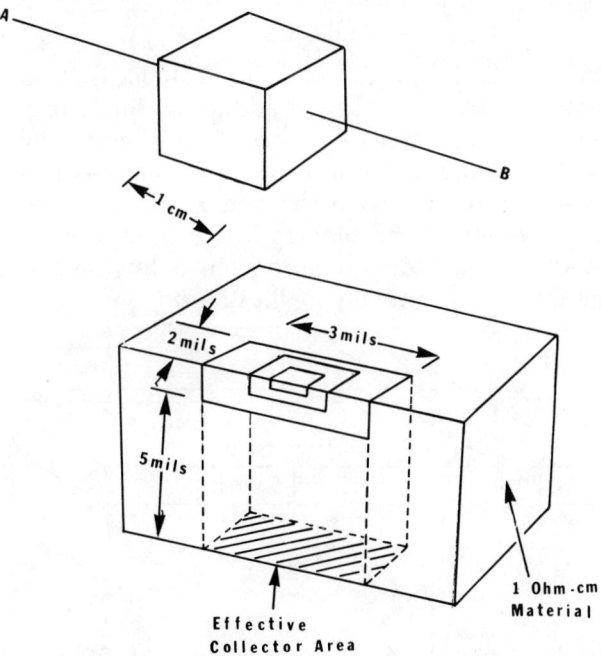

**2.5** What is the series collector resistance if 0.001 Ω-cm material is used for 4 of the 5 mils of wafer thickness? This is the technique used for conventional epitaxial transistors. Use the accompanying illustration.

## 22 Fabrication Materials

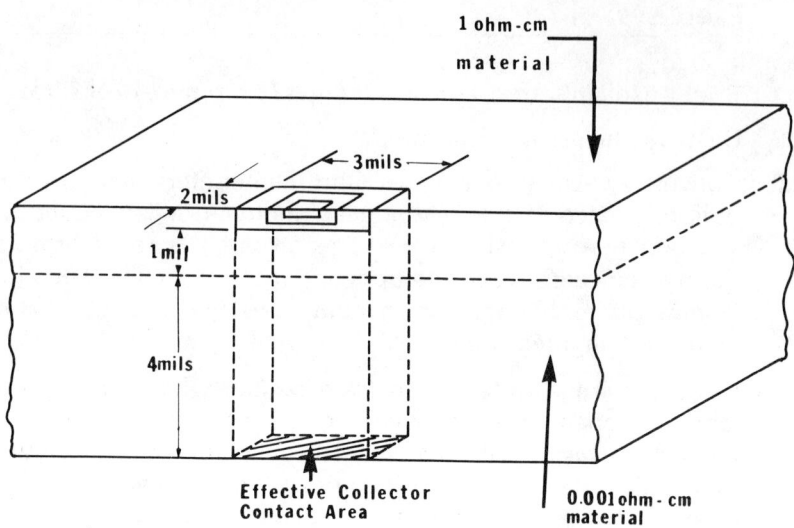

**2.6** With IC techniques, the collector contact is brought out to the top surface. What is the additional series resistance of a top collector contact ring 8 by 10 mils, and 1 mil wide (36 mils$^2$), assuming that the 0.001 Ω-cm material resistance can be neglected? The collector ring is metalized to eliminate contact resistance problems. This is now equivalent to an integrated transistor structure with a low-resistance buried layer. Use the accompanying illustration.

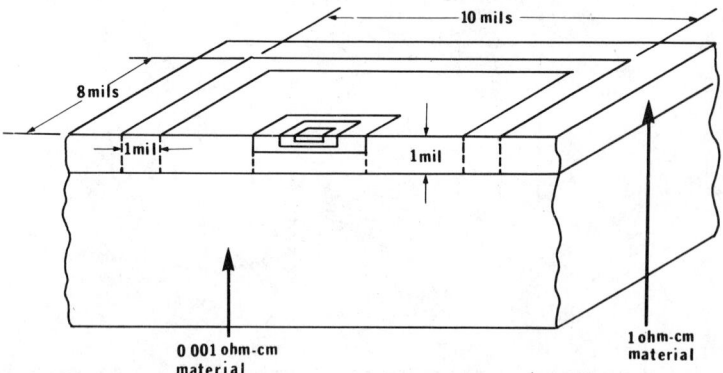

**2.7** During the probing of a slice, we find that the resistivity is 15 Ω, and the voltage of the circuit is 6 V. Assuming that the probes are 0.005 cm apart, what is the current?

chapter three
# Integrated Resistors

## 3.1 Silicon Resistors

The various components in an IC must be compatible with each other during fabrication. Furthermore, a rather desirable property of both active and passive components is that fabrication of the passive components requires no additional processing to that required for active components.

Silicon resistors appear to fall well within these requirements. As an example, observe the transistor of Fig. 3.1a and its integrated planar equivalent in Fig. 3.1b. Following the same processing steps, we can fabricate a transistor with a collector resistor by adding an ohmic contact. The resistance of the bulk silicon between parts b and c in Fig. 3.2 is the $R_c$ of the schematic; during actual fabrication, it would involve only the exchange of one or two masks and no additional steps.

As shown by this example, resistors are easy to fabricate. The quality of a silicon resistor must be determined not only by the usual criteria of resistor quality, such as tolerance in value, temperature coefficient of resistance, load-life stability, and noise, but also by criteria peculiar to functional electronic blocks, such as isolation, coupling, power dissipation, size, and circuit function.

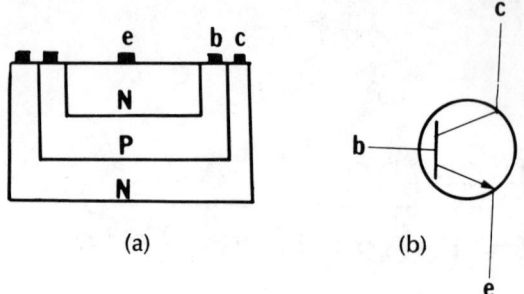

Figure 3.1  Planar n-p-n transistor and its equivalent symbol.

## 3.2 Diffused Resistors

One method of obtaining a compatible resistor is by diffusion. The resistor is obtained by diffusing a thin layer of *p*-type or *n*-type material. The resistance of this layer will depend on the concentration profile of the impurity in the diffused material, the depth of diffusion, and the length-to-width ratio of the diffused area. For uniformly doped bulk semiconductor material, the end-to-end resistance $R$ is given by

$$R = \frac{l}{tw}$$

where

$R$ = resitivity of the material (ohm-cm)
$l$ = length of the material (cm)
$w$ = width of the material (cm)
$t$ = thickness of the material (cm)

With diffused resistors used in IC's, the diffusion depth is extremely small and relatively constant. The resistance value of the diffused area stated in terms of the sheet resistance ($R_s$) of the material, measured in ohms per square, and the $l/w$ ratio of the diffused area is

$$R = \left(\frac{t}{t}\right)\frac{l}{w} = (R_s)\frac{l}{w}$$

When silicon is used, we can take advantage of the protective coating of silicon dioxide, which may be grown on the surface of the crystal.

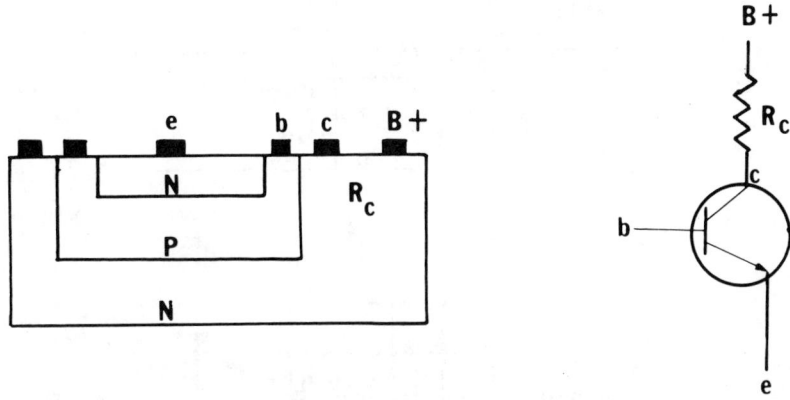

Figure 3.2  Planar n-p-n transistor with collector resistor, and equivalent symbol.

This dioxide coating acts as both a mask against impurity diffusion and as an insulating and passivity coating for the junctions. Figure 3.3 illustrates the construction of a typical unit.

As shown in Fig. 3.3a, an area, the length and width of which partially determines the value of the resistor, is etched through the silicon dioxide. The unit is then placed in a furnace for a boron diffusion to form a $p$-type layer, several micrometers deep, as shown in

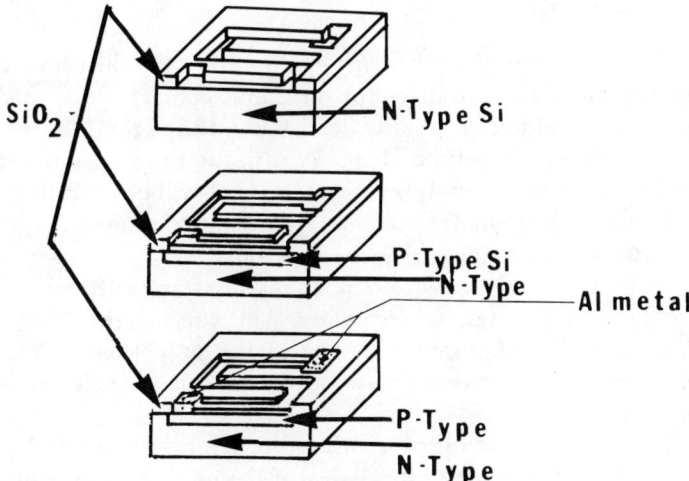

Figure 3.3  Fabrication steps of a diffused resistor.

**26  Integrated Resistors**

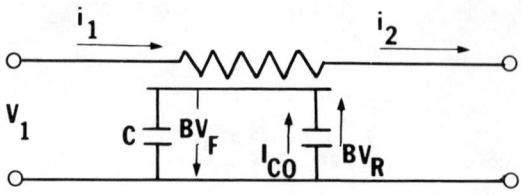

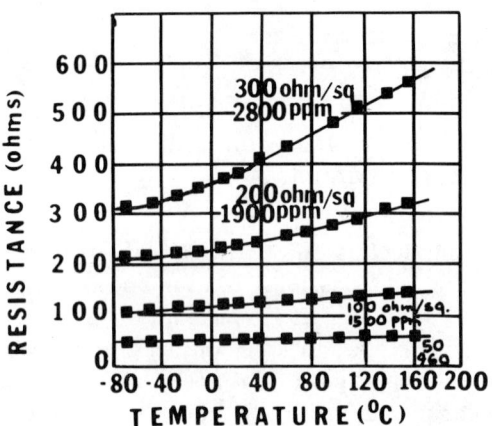

Figure 3.4  Equivalent circuit of a diffused resistor, and various resistor values plotted as a function of temperature.

Fig. 3.3b. The silicon dioxide forms again during the diffusion, and is then etched again to form holes for the ohmic contact areas, as shown in Fig. 3.3c. Aluminum or gold is deposited into the holes to form the ohmic contact to the $p$-type layer. Wire leads may be bonded into the metalized areas by bonding techniques that will be described later. Typical values of resistance obtained by this method may range from 10 Ω to 100 kΩ. Tolerances may be controlled to ±10 percent in the diffusion process. When two or more resistors are diffused into the same substrate, the resistance values will vary slightly, even for identical pattern configuration, owing to the properties of the substrate. Normally, however, these variations can be held to within ±3 percent.

The equivalent circuit of a diffused resistor is shown in Fig. 3.4. A diode and the distributed capacitance of the $pn$ junction are available when a contact is made to the substrate material. Some of the values of the parameters that are indicated in Fig. 3.4 are as follows:

$BV_F = 0.5$ V
$BV_R = 50$ V
$I_{CO} = 10$ nanoamperes (nA)

where

$BV_F$ = forward voltage drop across the junction
$BV_R$ = reverse breakdown voltage of junction
$I_{CO}$ = leakage current of junction

The value and variation of the capacitance are discussed further later. These parasitics must be taken into account, as will be shown, when the substrate becomes part of an integrated circuit.

The following parameters are usually considered, in addition to the preceding values, to describe a diffused resistor: (1) frequency effects, (2) parasitic capacitance, (3) temperature coefficients, and (4) maximum ratings.

Figure 3.5 illustrates, in terms of $h$ parameters, some of the characteristics of a diffused 5-k$\Omega$ resistor as a function of frequency. These results were obtained with the substrate as the common ground and with the resistor terminals as the input and output. The plot of

$$h_{21} = i_2/i_1$$

where

$i_1$ = input current
$i_2$ = output current

illustrates the transfer characteristics, whereas the plot of

$$h_{11} = v_1/i_1$$

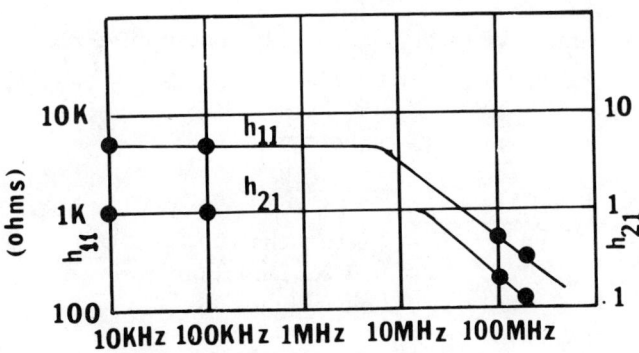

Figure 3.5 Characteristics of a diffused 5 kilometer resistor as a function of frequency.

where $v_1$ = input voltage, illustrates the actual values of resistance as a function of frequency. As shown, these measurements indicate useful action up to nearly 10 MHz. However, by isolating the substrate, the parasitic capacitance effect is reduced and the frequency range is extended.

For resistors of this type, the distributed junction capacitance is a function of both the impurity concentration of the substrate and the voltage across the *pn* junction. As an example, a diffused resistor of 100 Ω/square, diffused 3 μm deep into 0.5 Ω-cm silicon, will have approximately 0.13 pF/mil² at 1-V reverse bias. The value of this capacitance also closely follows the theoretical form for a reverse-biased graded junction of

$$C = KV^{-1/3}$$

where $K = 18 \times 10^{-12}$. Thus the effective junction capacitance may be governed by a reverse-biased voltage placed on the resistor substrate.

The temperature coefficient is another variable that should be considered in the application of these devices to IC's. One of the most critical parameters affecting the value of the temperature coefficient is the surface impurity concentration of the diffused area. The total resistance $R$ of the units is determined primarily by $\rho$ by

$$R = \frac{\rho_{ave} l}{A}$$

where

$l$ = length of the diffused area
$A$ = width times depth of the junction
$\rho_{ave}$ = average value of the resistivity in the diffused area

From semiconductor theory, in a highly concentrated *p*-type layer, $\rho$ is given by

$$\rho = \frac{1}{q \mu_p p}$$

where

$q$ = electron charge
$p$ = hole concentration
$\mu_p$ = hole mobility

Or for *n* type,

$$\rho = \frac{1}{q N_{D_e}}$$

Thus, in the normal temperature range, $p$ is the primary temperature-dependent parameter in the determination of $R$. Referring to Fig. 3.4, we observe various resistor values plotted as a function of temperature. The top curve illustrates the effect of a low value of impurity concentration (approximately $10^{18}$ atoms/cm$^3$); the lower curves show the effect of higher concentrations. At 100 $\Omega$/square diffused approximately 3 $\mu$m deep, the temperature coefficient is approximately 1500 parts per million (ppm) per degree Celsius. Thus the temperature coefficient of the diffused resistor may be controlled to a great extent by the initial surface impurity concentration of the diffused area.

There are some inherent low-frequency limitations in the application of these devices, mainly power dissipation and maximum voltage ratings. The amount of power dissipation in the resistor is limited primarily by the heating effects on the material of the diffused layer. Excessive heating will result in a nonlinear current–voltage relationship.

Parasitic capacitance has been mentioned previously. The inherent parasitic junction capacitance of the device may be used to an advantage. In some applications, such as the coupling resistor of a flip flop, a small speed-up capacitor is usually placed in parallel with the resistor. By connecting the isolated substrate to one terminal of the diffused area, part of the junction capacitance will appear across the resistor.

## 3.3 Epitaxial Layer Resistors

An epitaxial-layer (epi) resistor is formed in an epitaxial layer of silicon upon a substrate of opposite-type-conductivity silicon. An example is shown in Fig. 3.6. The resistors are defined in area by masked diffusion using the same conductivity type of diffusant impurity as that of the substrate. These diffusions must penetrate through the epitaxial layer, as shown in Fig. 3.6b. Contacts are alloyed to the surface of the epitaxial region, and the spacing between them is used to compute the resistor value just as was done for the diffused resistor. The major difference between the two is that the impurity distribution of an epitaxial layer approaches a uniform distribution.

## 3.4 Vapor-Deposited Resistor

The vapor-deposited resistor is formed by the same process as the epi resistor, except that the silicon substrate now has a layer of silicon dioxide on it. The effect of the layer of SiO$_2$ is to prevent

**30   Integrated Resistors**

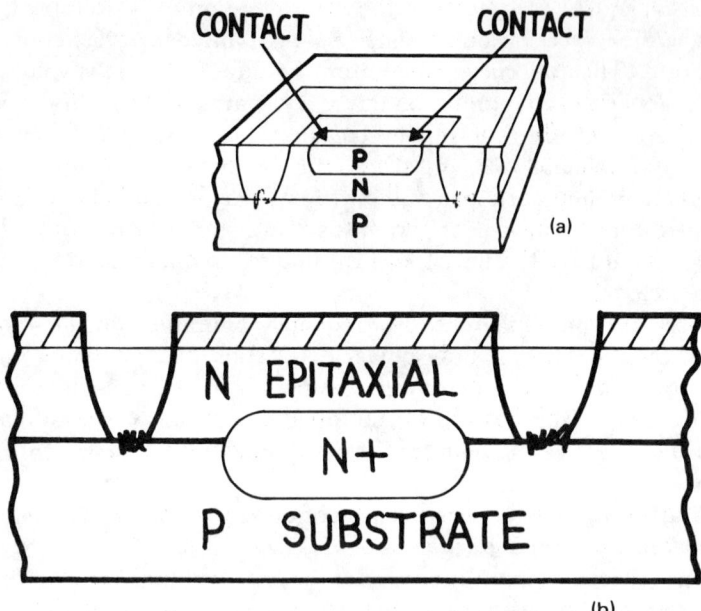

Figure 3.6

epitaxy and to cause a randomly oriented polycrystalline layer to form as the silicon atoms condense. The resistor geometry must be controlled by a subsequent operation such as etching. Figure 3.7 illustrates two steps in the fabrication of such a vapor-deposited resistor in silicon.

## 3.5 Thin-Film Resistors

The electrical properties of thin films differ from those of bulk materials of the same composition. The resistivity is usually higher and depends upon the film preparation and substrate properties. The effects of impurities upon resistivity are much less pronounced than for silicon resistors. Film resistors have been made of many types of materials and deposited by various techniques.

## 3.6 Practical Design

As can be concluded from the previous discussions, the most common method is to utilize the base diffusion cycle to form a resistor. With a diffused resistor, the depth is extremely small (approxi-

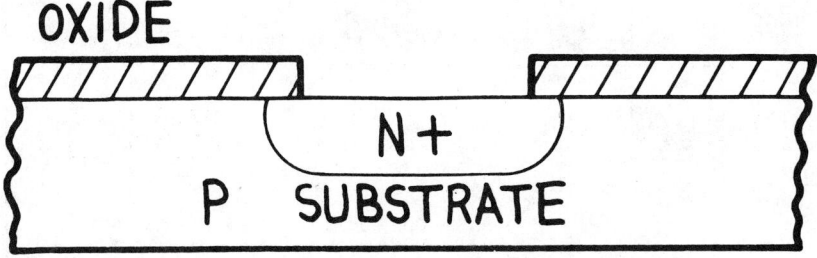

Figure 3.7

mately 3 μm) and is relatively constant. Therefore, we arrive at the equation of Section 3.2.

Figure 3.8 shows the mask layout of a 600-Ω diffused resistor. The total resistance is determined by the 2- by 6-mil area of the 200 Ω/square, p-type base diffusion. The edge of the n-type region must be approximately 2 mils (50 μm) away from the resistor area to allow for the spreading of the p-channel isolation diffusion. Due to the fact that the contact area cannot extend completely across the diffused area without shorting the junction, there is added resistance in the contact area. This is an example of a resistor end effect. Example 2 of Fig. 3.9 gives the correction factor for this type of resistor. The actual effective length is greater by a factor of 2 × 0.14, or the total resistance is

$$R = \frac{R_s}{2}[1 + 2(0.14)]$$
$$= \frac{200}{2}[6 + 2(0.14)]$$
$$= 628 \, \Omega$$

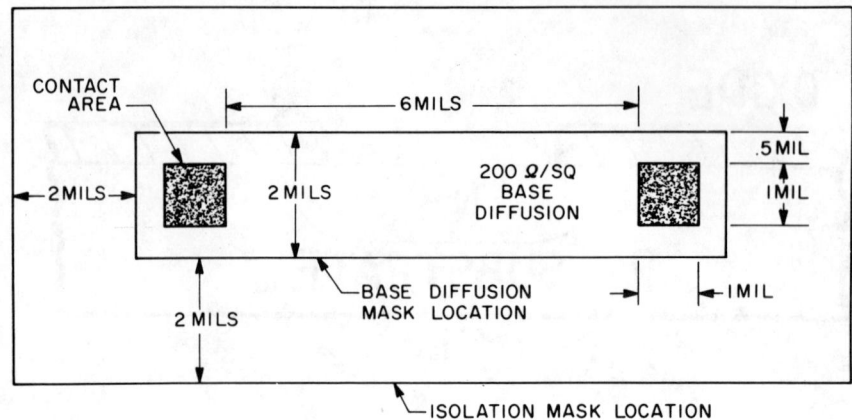

Figure 3.8

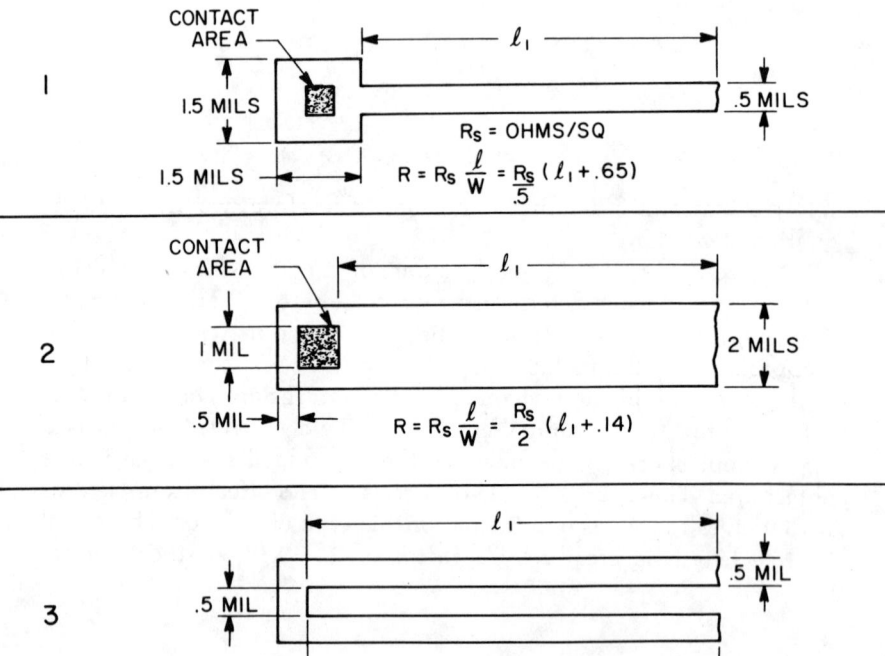

Figure 3.9

# Exercises

**3.1** Give a brief description of the differences in the fabrication of resistors.

**3.2** Describe the epi resistor.

**3.3** A silicon bar is doped with phosphorus at a concentration of $N_D = 10^{15}$ cm$^{-3}$. The mobility of electrons is 1400 cm$^2$/V-sec, and the magnitude of the electronic charge is $1.6 \times 10^{-19}$ coulomb (C). Find the resistivity of the doped semiconductor.

**3.4** Sketch and calculate the total resistance of a diffused resistor 4 by 7 mils when the diffused material is 160 $\Omega$/square. Take into account the end effects.

**3.5** Describe the vapor-deposited resistor.

**3.6** Find the distributed junction capacitance of a diffused resistor when the voltage is 0.7 V.

**3.7** Describe the action of the speed-up capacitor in a flip flop.

## chapter four
# Integrated Capacitors

## 4.1 Introduction

The capacitor is one of the most critical components in the integrated circuit, and its fabrication is the result of many years of research. Capacitors occupy a considerable portion of the IC die, and are avoided whenever possible, particularly if they are of large values. There are several types of integrated capacitors, as described next.

## 4.2 Diffused Capacitors

Two of the early methods of fabricating a capacitor are shown in Fig. 4.1a, b, and c. The first is the emitter–base junction method, which yields reasonable capacitances (1 pF/mil$^2$ at zero bias), but offers a very high series resistance because the base diffusion region under the emitter is very thin and of very high resistivity (about 3 k$\Omega$/square). This type of capacitor is almost useless.

The second type of capacitor can be fabricated by the collector–base junction method, and this type of capacitor offers low series resistances, provided the contact extends over most of the base region. However, the obtained capacitance is very small, and any at-

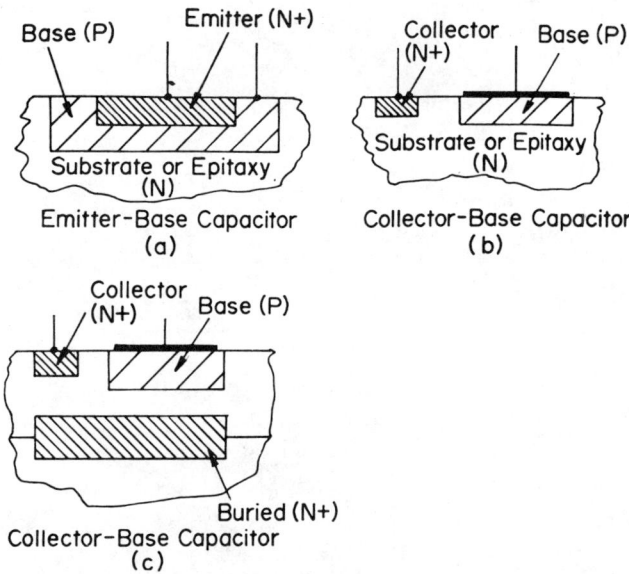

Figure 4.1 Monolithic capacitors.

tempt to increase the capacitance will conflict with the general requirement for low transistor collector–base capacitance.

## 4.3 Epitaxial Capacitors

When the epitaxial buried $n+$ process was adopted, methods were sought to provide high-value capacitors suitable for decoupling points at positive or negative potentials with respect to ground. Figure 4.2 illustrates an epitaxial capacitor. The buried $n+$ layer, normally used to reduce the transistor collector series resistance, is used here to form the lower plate of the capacitor. This layer is diffused into the original $p$-doped substrate material to provide the bottom plate of the capacitor. The layer is arsenic-doped, has a surface concentration of approximately $10^{21}$ atoms/cm$^3$, and a sheet resistance of about 6 $\Omega$/square.

The next step is to grow an epitaxial layer of 0.68 $\Omega$-cm onto the $p$-doped substrate. The isolating diffusion is then used to form the center plate of the capacitor. This diffusion is of boron and has a surface concentration of approximately $5 \times 10^{19}$ atoms/cm$^3$. Since the diffusion is deep (i.e., sufficient to diffuse completely through the epitaxial layer), it is also of a low resistance (about 15 $\Omega$/square).

Next the emitter $n+$ diffusion is diffused so as to overlie the

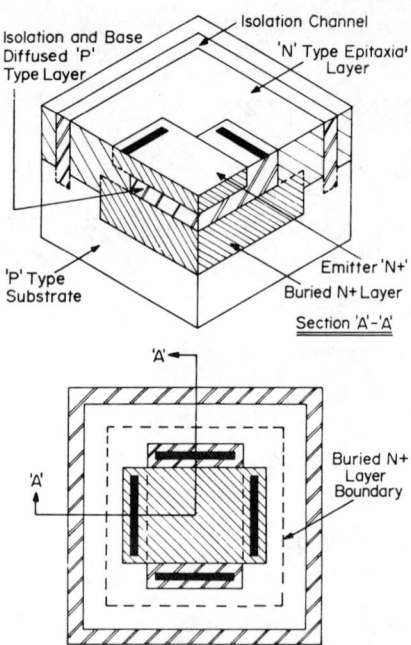

Figure 4.2  Epitaxial capacitor.

isolation diffusion to provide the upper plate of the capacitor. The surface $n+$ diffusion is connected to the buried $n+$ diffusion by means of the overlay, conduction being via the epitaxial $n$-type layer. The center, isolation-diffused plate is extended beyond the upper plate along part of its periphery to enable contact to be made from the surface of the device. This type of capacitor, unlike the diffused capacitor, is not voltage dependent.

## 4.4  Design of Diffused Capacitors

The diffused capacitor offers the advantage of easy fabrication. It can be incorporated into an integrated circuit without new processing steps, and this greatly affects cost, reliability, and yield. In the present state of the art, a diffused junction capacitor does not otherwise exhibit any particular advantages that would make it a reasonable choice for a majority of applications. Exceptions to this statement are those applications in which advantage can be taken of the voltage dependence that the diffused capacitors present.

There are a number of reasons why a junction capacitor might not be chosen for a given application. The dependence of voltage upon voltage can be objectionable, as in the case of coupling, when the capacitance variation with the ac signal may introduce intolerable distortion.

An additional important consideration is the area to be occupied on the chip in order to obtain the desired value of capacitance. As a typical example, the capacitance associated with an emitter is $10^5$ pF/cm$^2$, whereas that associated with a collector junction is less (typically $10^4$ pF/cm$^2$).

Practical capacitance values require large space, and, since the manufacturing yield is directly related to the areas of the device, the area required by capacitors is expensive. The practical limitation of junction capacitors to those junctions that are utilized for other structures in the silicon chip dictates a very limited range of values. For example, utilizing a 100-mil$^2$ chip, and assuming that one half of this area is available for a capacitor, the capacitance of a typical emitter junction covering this area is 3200 pF.

The voltage breakdown of an emitter junction is about 6 V. In some circuits, this value may be acceptable; in others, it may be intolerable. In the latter case, one must design around the lower capacitance per unit area of the collector junction.

For isolation of the *pn* junction capacitor from the substrate, small parasitic capacitance and high resistance between the capacitor and substrate are desirable. Figure 4.3 portrays a cross section of a *pn* junction capacitor employing the emitter–base diffusion; the equivalent circuit for the capacitor is shown in Fig. 4.4. The transistor enclosed in the square represents the base–collector, collector–substrate junctions, and the possibility of transistor action between these junctions. Consequently, transistor action between the junctions depends upon the separation distance of the junctions.

Transistor action between the junctions is obviously a problem. Shorting the base to the collector will obviate this problem, and the

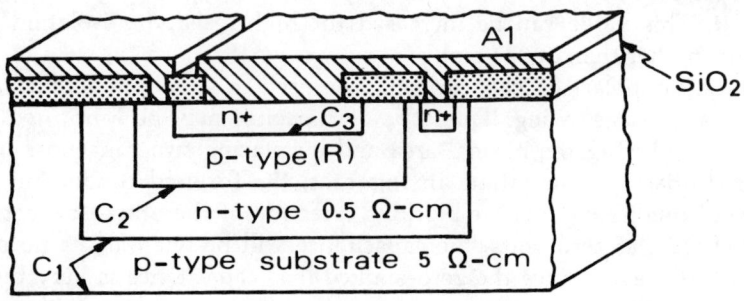

Figure 4.3  Junction capacitor.

## 38 Integrated Capacitors

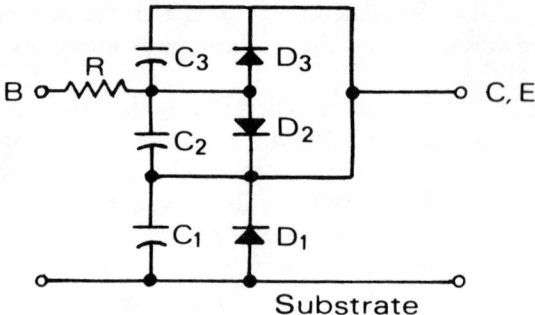

Figure 4.4   Equivalent circuit.

parasitic capacitance will then be due to the collector–substrate junction. In this case, the reverse bias of the emitter–base junction can be used to reverse bias the collector–substrate junction. However, for applications in which transistor action due to parasitics is unlikely to occur, a floating collector can be used. In this case, the parasitic capacitance consists of the base–collector junction capacitor in series with the capacitance of the collector–substrate junction.

If the substrate is grounded, either a positive or negative voltage (with respect to ground) may be applied to the base contact, and one of the parasitic junctions will be reverse biased, reducing the capacitance of that junction.

When the resistivity of the collector region increases, the parasitic capacitance decreases. However, the practical limit will be determined by the transistor design. Typical practical values are one tenth to one twentieth of the value of the emitter–base capacitance.

The isolation resistance is determined by the resistance of the reverse-biased junction and by the resistivity and geometry of the isolation regions. If the isolation between elements is a diffused region in an epitaxial layer, the major portion of the isolation resistance will be that of a reverse-biased junction, which is typically greater than $10^6$ Ω. This value can be increased for high-resistivity isolation regions (higher than 100 Ω-cm).

A nonpolarized capacitor may be used in situations requiring a large ac voltage swing. This type of capacitor may be fabricated by twice employing the normal area and by placing two capacitors in a back-to-back configuration. In this case, the forward conduction region of one junction is the blocking direction of the other. It must be noted that, at zero volts, the capacitance will be one-half its normal value. However, since the zero-applied-bias capacitance is very high, it may assist in the reduction of the capacitance variation with voltage.

The practical design of a junction capacitor is fairly simple, once the particular value and the type of capacitor have been chosen. The area that the capacitor may occupy can be calculated on the basis of the theoretical capacitance per unit area, or, preferably, upon empirical measurements made for the particular junction that will be employed.

When the area to be occupied is specified, the shape of the capacitor, and, to some extent, its contact configuration are determined by the overall topology of the IC. The contact area should be of such construction and dimensions as to give a minimum series resistance, and, for the emitter–base junction capacitor, need only cover a portion of the low-resistivity emitter region.

## 4.5 MOS Capacitors

Metallic-oxide semiconductor capacitors have proved to be very useful. They are primarily used as IC capacitors. Visually, the $n$-type epitaxial layer with a shallow, heavily doped, $n+$ type of diffusion is used. This type of region can be isolated from the other circuit elements either by a channel diffusion around the $n$ region or a $p$-type base diffusion under the heavily doped $n+$ type of diffusion.

This capacitor provides rather small capacitance (0.1–0.8 pF/mil$^2$), but also offers a very low series resistance and is not voltage dependent. A typical MOS capacitor appears in Fig. 4.5.

The equivalent circuit of an MOS capacitor is shown in Fig. 4.6. $R_s$ is an equivalent series resistance, which includes the plate and lead resistance of the silicon and metal electrodes; $R_p$ is the equivalent leakage resistance of the MOS capacitor; and $R_s'$ is associated with the time constant for the surface slates.

The above equivalent circuit can be greatly simplified if the capacitance contribution which is created by the surface slates is neglected. Figure 4.6 portrays a simplified version of the MOS equiv-

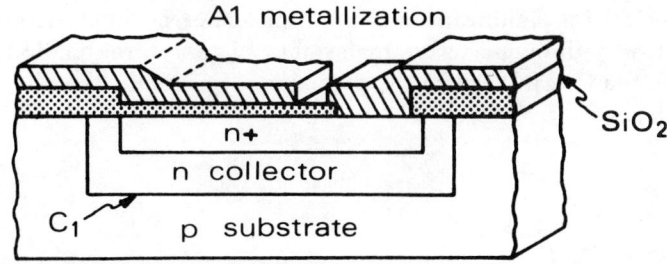

Figure 4.5 MOS capacitor.

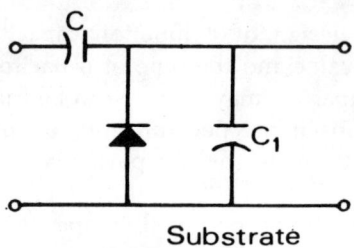

Figure 4.6   Equivalent circuit.

alent circuit. It is worth noting here that the equivalent circuit represents an ideal MOS capacitor.

## 4.6 Design of MOS Capacitors

Commonly used oxide thicknesses in the fabrication of an MOS capacitor vary from 500 to 1000 Å. The capacitance per unit area thus varies from 0.2 to 0.4 pF/mil$^2$, with a rated breakdown voltage of 20–40 V. The breakdown of these capacitors is destructive. If the dielectric breakdown is exceeded, the aluminum electrode shoots through to the lower electrode.

An additional property of the MOS capacitors is their appreciable parasitic capacitance. This capacitance is created by the *pn* junction formed between the lower silicon electrode and the isolation region, and is the same as that present in diffused capacitors. The capacitance of a particular MOS capacitor per unit area is larger, and thus the ratio of the desired capacitance to parasitic capacitance is higher. At 6-V reverse bias, the improvement of the MOS capacitor to the diffused capacitor is 0.3/0.085 or 3.5. It is worth noting, however, that the parasitic capacitance is harmful only when the capacitor is used for coupling, and not harmful when the capacitor is used in a bypass application. The calculation of parasitic capacitance of the MOS and diffused capacitors is more complicated by the sidewall capacitance of the isolation area. This capacitance is significant because the area is significant, and the capacitance per unit area of the sidewalls is high owing to the high impurity concentration. The sidewall capacitance can become the predominant parasitic on small-value capacitances.

## Exercises

**4.1** Outline the differences between junction and MOS capacitors.

**4.2** What are the disadvantages of integrated capacitors in a circuit?

**4.3** Describe briefly an epi capacitor.

**4.4** Where should a nonpolarized capacitor be used?

**4.5** During fabrication of an MOS capacitor, the dielectric breakdown is exceeded. What may we expect of the circuit?

**4.6** Provide an equivalent circuit for a diffused capacitor.

**4.7** If a capacitor is to occupy one third of a 60-mil² chip, what is the approximate capacitance?

**4.8** An integrated circuit calls for a capacitor that should not be voltage dependent. Which type of capacitor should be used?

chapter five
# Integrated Diodes and Transistors

## 5.1 Basic Diode Concepts

Integrated diodes are usually formed from the basic transistor structure. The several methods by which a transistor can be fabricated to operate as a diode are shown in Fig. 5.1. Transistors in IC's are cheap, and are often the basic structure for diodes and other components in an IC.

Of the five different methods of producing a diode, two are the most acceptable, the low reverse voltage type (emitter–base junction), and the medium voltage type (collector–base junction). The shorted collector–base configuration ($V_{CB} = 0$) also offers the fastest recovery time, since in it minority carriers are stored only in the base; in all other configurations, excess carrier densities are established in both base and collector regions during conduction.

An important consideration with respect to speed is the transition capacitance of the diode. Typical emitter junction capacitance is about $5 \times 10^4$ pF/cm$^2$; that of the collector junction is about $10^4$ pF/cm$^2$. Collector area may be $5 \times 10^{-5}$ cm$^2$, as compared to $10^{-5}$ cm$^2$ for the emitter, so the capacities are about equal. The isolation junction capacitance is less owing to the higher resistivity material

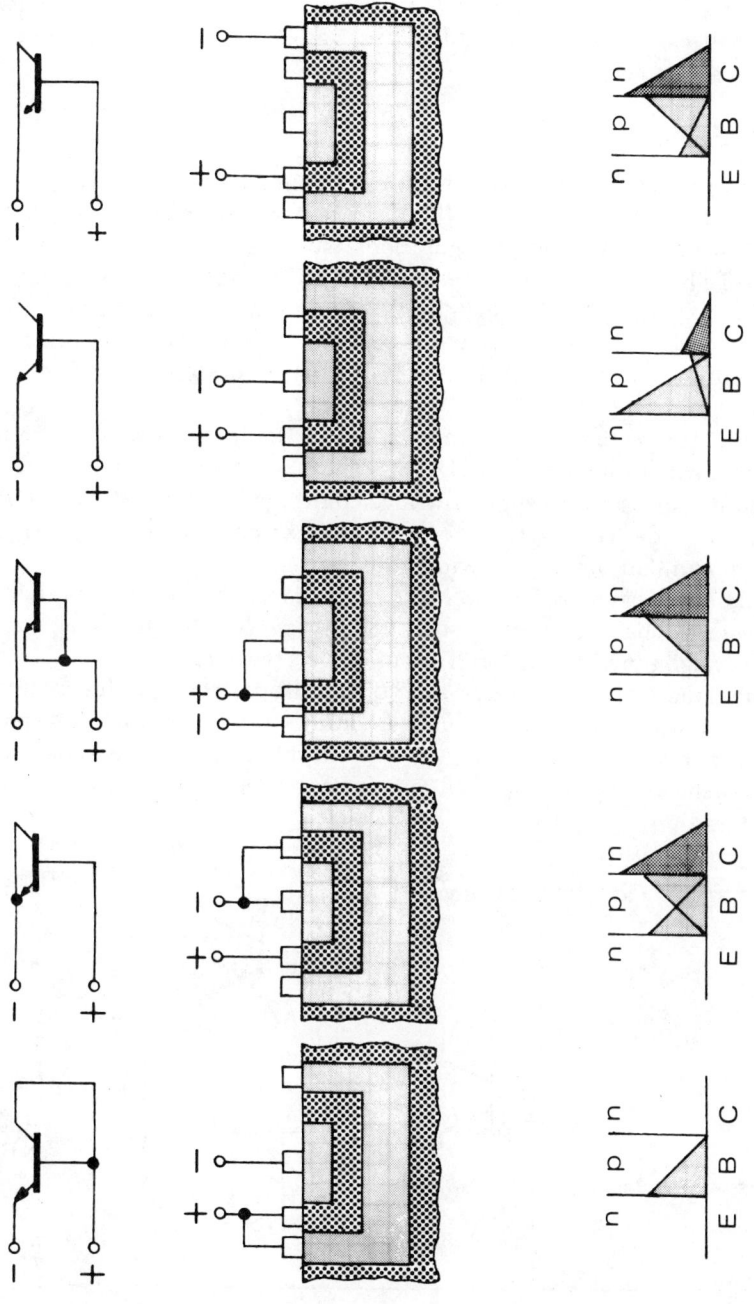

Figure 5.1 Diode configurations.

bounding it. A typical value is $5 \times 10^3$ pF/cm², but this, as are other capacitances, is voltage dependent and varies with the doping levels used in the structure.

## 5.2 Diode Applications

In the past, diodes have been used abundantly in the design of logic circuits. However, with the development and growth of faster and more reliable logic configurations, such as transistor–transistor logic (TTL) and emitter-coupled logic (ECL), diode–transistor logic (DTL) is getting to be something of the past. Nevertheless, the emitter–base Zener diode is at present found in a variety of devices, and transistors connected as diodes are used for temperature compensation.

An interesting application of diodes is the controlled-output current switch illustrated in Fig. 5.2. The basic function of this integrated design is to accept a low-level pulse and convert it to a 1-A pulse. It is desired that the output not exceed 1 A by any large amount and that the device switch rapidly.

In the design of Fig. 5.2, the diodes are shorted collector–base transistors. Their function is such that, at low inputs, the diode current is negligible, and the full input drives the Darlington pair into conduction. At higher inputs, the voltage drop across the transistors is such that the diodes conduct, shunting any excess input signal above that required to give 1 A at the output. Temperature stabilization is achieved by the matched temperature coefficients of two of the

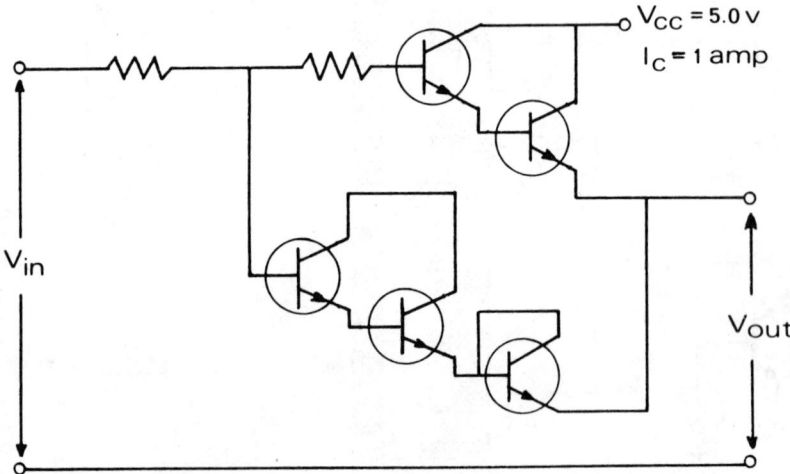

Figure 5.2 Current switch.

Diode Applications  45

diodes and the transistors, and the fact that the forward-voltage drop of the third diode decreases with increasing temperature while the Darlington beta increases. The result is compensation.

As mentioned, diodes are used for temperature compensation, and another example is as follows: consider the amplifier of Fig. 5.3. It is assumed that this differential amplifier is designed for impedance matching and isolation purposes, and is required to have an output impedance of less than 1 k$\Omega$, with an input impedance larger than 10 megohms (m$\Omega$). The voltage gain, for negative input voltages between 0 and 6.4, is to be unity, with a zero signal dc offset voltage of less than 5 mV.

A particular problem with differential amplifiers of this type is the drift in the dc characteristics of the amplifier components due to temperature variations. Integrated fabrication minimizes the problem because it offers closer thermal coupling between the separate component regions. Furthermore, input and output voltage equalization is also enhanced by the similarity of input and output junctions provided by the prudent use of compensating diodes.

In the basic schematic, $Q1$ and $Q2$ are *npn* transistors with the same emitter resistance, $R1$, a typical differential amplifier configuration. The output from the collector of $Q1$ is the input to $Q3$, and the collector output of $Q3$ is also the output of the amplifier and the input of $Q2$. The base–emitter voltage drops in $Q1$ and $Q2$ should be equal, provided the base currents are equal. The input and output voltages are then identical.

An improved version of the circuit is shown in Fig. 5.4. $Q1$ and

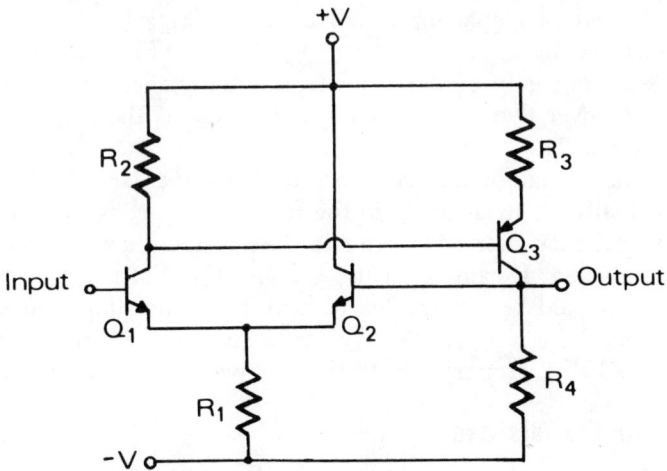

Figure 5.3  Basic buffer amplifier.

## 46 Integrated Diodes and Transistors

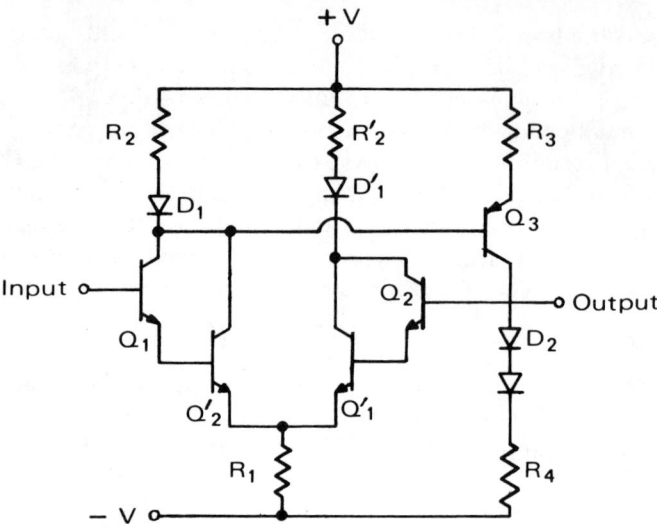

Figure 5.4  Improved version of buffer amplifier.

Q2 have been replaced by Darlington amplifiers to provide an increased input resistance. Diode $D1$ has been placed in series with the collector resistor $R2$, not only offsetting the base–emitter voltage of $Q3$, but also providing for the same temperature dependence of this voltage. The double-diode structure $D2$, in series with $R4$, provides both voltage and temperature compensation for the equivalent emitter junctions of $Q2$ and $Q2'$.

In the improved design, the dc currents and voltages are carefully balanced for equal input and output voltage levels. The input impedance of the circuit is effectively $R1$ multiplied by twice the effective current gain of the input Darlington stage. The output impedance is lower than the value of $R4$ because of the negative feedback action of $R1$.

In the actual integrated fabrication of the circuit, diode $D2$ would actually be constructed in the form of a Darlington pair, with collector and base shorted to provide close temperature compensation of the equivalent Darlington pairs $Q1$ and $Q2$. Furthermore, diodes $D1$ and $D1'$ would be constructed to have junctions similar to that of the emitter of the *pnp* transistor, with the actual area and peripheral dimensions being proportional to the relative current densities.

## 5.3 Bipolar Transistors

The integrated transistor is the most widely used device in an IC, and therefore is entitled to a thorough discussion in the text. The operation of the bipolar transistor depends on the injection of minority carriers at a forward-biased emitter–base junction, followed by diffusion across the base region, and the collection of minority carriers at the reverse-biased collector junction. The injected minority carriers constitute a nonequilibrium charge, which results in the recombination of excess hole–electron pairs in the base. This recombination may occur either in the bulk material or at the surface. The minority carriers are supplied from the emitter, but the recombining majority carriers must be supplied to the base, and this constitutes one of the major components of the base current.

The preceding process greatly contributes to the dc emitter and collector currents. The injection of minority carriers from the base into the emitter is often referred to as *emitter inefficiency*. This current is minimized by heavily doping the emitter as compared to the base.

The contribution from the collector junction is usually negligible when the collector–base junction is reverse biased, but becomes an important factor when the junction goes into saturation, that is, when the collector–base junction is forward biased.

The recombination of excess electron–hole pairs in the junction depletion region is important for low-current operation. It has a different dependence upon the emitter–base voltage from the emitter inefficiency and the bulk recombination current, and results in a decreasing beta at low currents.

Figure 5.5 clearly illustrates the various paths of electron current in an *npn* transistor. When the transistor is forward biased (EB junction forward biased, CB junction reverse biased), the arrows are in the direction of the electron flow, with the exception of arrows 7 and 8, which go in the opposite direction.

The relative importance of the various current paths depends to some extent upon the transistor operating conditions. For example, component 7 is negligible under normal linear operating conditions, as compared to the total collector current. However, under saturation, this component becomes of significant importance to the collector current.

The current flow at each junction can be written as the sum of (1) the minority diffusion and drift currents on both sides of the junction, (2) the generation–recombination currents from the depletion region and from the surface near the junction, and (3) the displacement currents flowing to the boundaries of the junction depletion region.

**48    Integrated Diodes and Transistors**

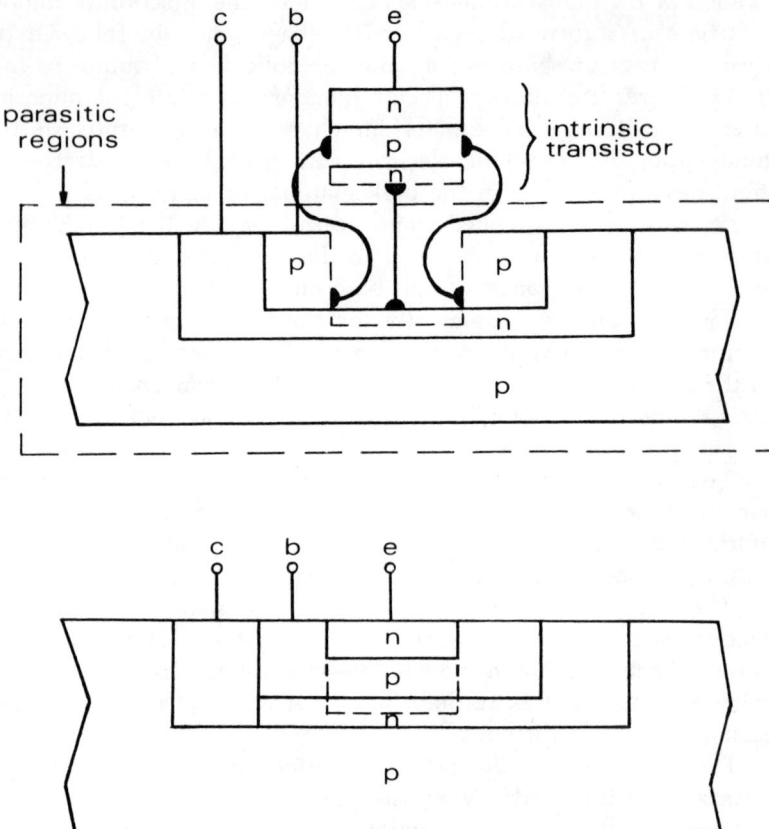

Figure 5.5  Regions of transistor structure contributing to parasitic elements.

## 5.4 Parasitic Elements

The major current components discussed in the preceding sections are inherently associated with diffused transistors. Similarly, there are resistive- and capacitive-type elements (parasitic), which, although not inherently associated with the minority carrier transport between emitter and collector, must be accounted for in the description of terminal characteristics of devices.

Figure 5.6a illustrates a typical integrated transistor with an isolation junction; Fig. 5.6b shows the division of the transistor into intrinsic and parasitic regions. The intrinsic transistor is taken as the *npn* region directly under the emitter junction, whereas the parasitic

## Parasitic Elements 49

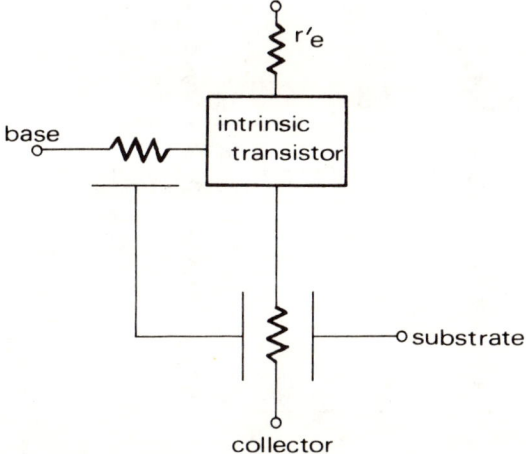

Figure 5.6  Distributed R-C representation of parasitic regions.

regions include the regions of the base and collector not directly beneath the emitter junction, substrate, the isolated junction of the transistor, and any resistance associated with the collector region.

Any capacitance and resistance between the external terminals and the internal intrinsic transistor regions is contributed to the parasitic regions. Figure 5.7 is a schematic representation of the parasitic regions, which are shown as an *RC* network. What the illustration does not show is the parasitic *pnp* transistor action that can actually occur between the substrate, collector, and base region.

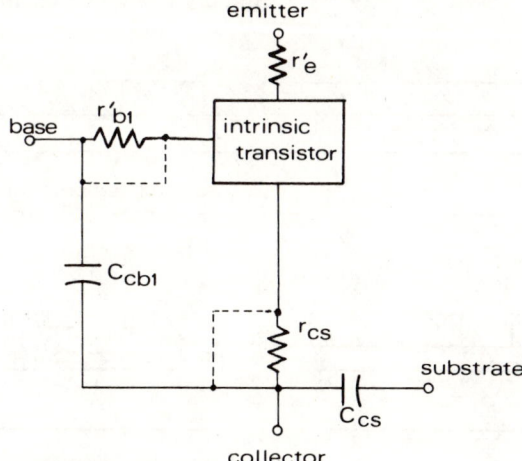

Figure 5.7  Simple approximation to parasitic regions.

**50** Integrated Diodes and Transistors

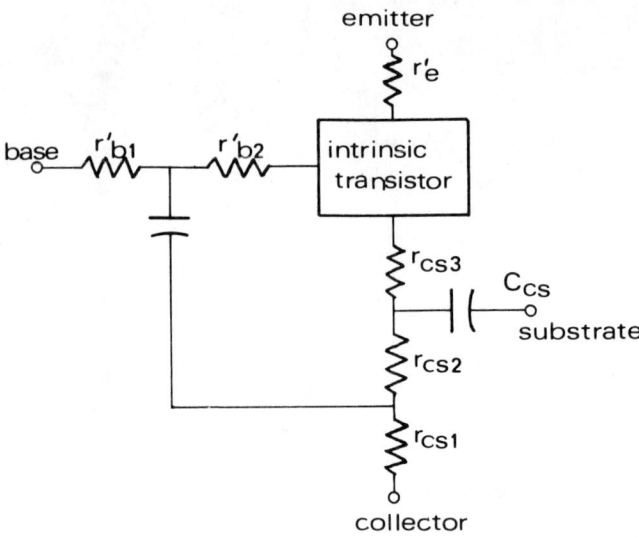

Figure 5.8 High frequency approximation to parasitic regions.

Figure 5.8 is an approximation of the parasitic regions at low frequencies. In this case, the regions merely provide resistances in series with the base and collector contacts. Figure 5.9 is a more accurate approximation of the parasitic regions at high frequencies during which junction capacitance obviously becomes more important.

An accurate approximation of the parasitic elements, both capacitive and resistive, may be a difficult task. While, in some cases,

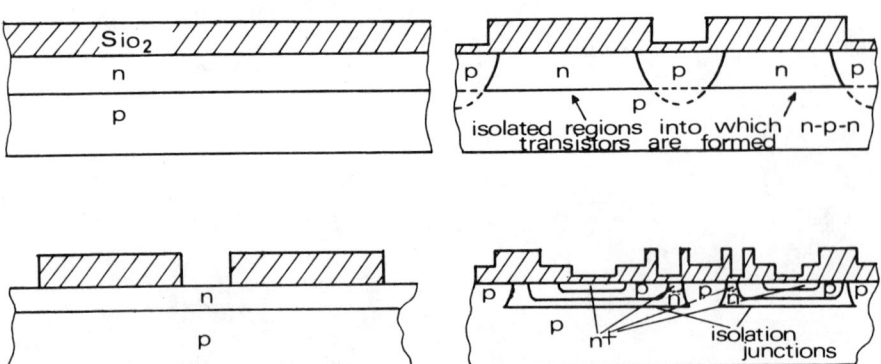

Figure 5.9 Epitaxial technique.

capacitance can be easily calculated, resistance is more complex because of the different geometries and contact positions possible with transistors.

## 5.5 Integrated-Transistor Structures

The epitaxial diffused transistor is most frequently used in IC's. The process sequence is shown in Fig. 5.10. The first step is an oxidized epitaxial wafer with a typical layer thickness of 20 μm. Isolated n-type regions surrounded by p-type regions are then formed by etching the isolation pattern into the oxide (Fig. 5.9b), and by diffusing p-type regions through the epitaxial layer to the p-type substrate (Fig. 5.9c). The final product appears in Fig. 5.9d.

Since the transistor collector region (n-type epi layer) can be controlled independently of either the substrate or the base region resistivity, the epitaxial process may be used to vary the epi layer

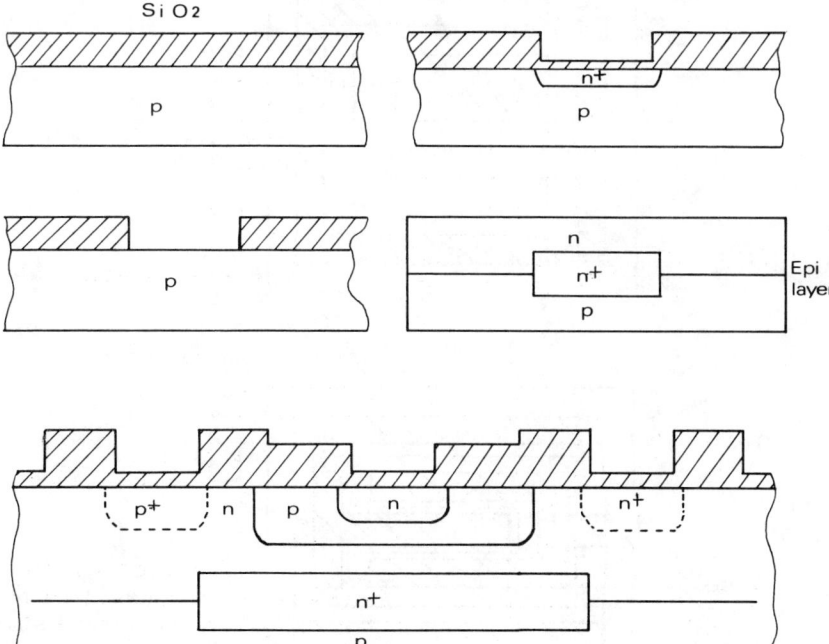

Figure 5.10  Buried layer technique.

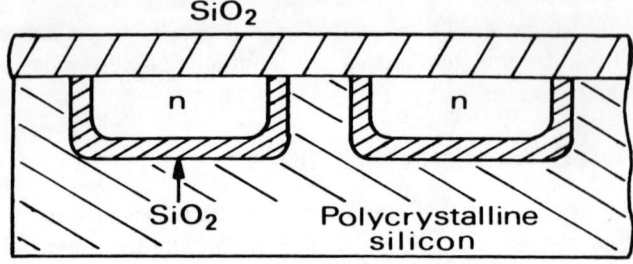

Figure 5.11 Oxide isolation technique.

resistivity for optimum collector resistance and isolation junction capacitance, and this at no expense to the basic transistor characteristics. Typical substrate resistivity, with an epi layer of 0.5 $\Omega$/cm, is 10 $\Omega$-cm.

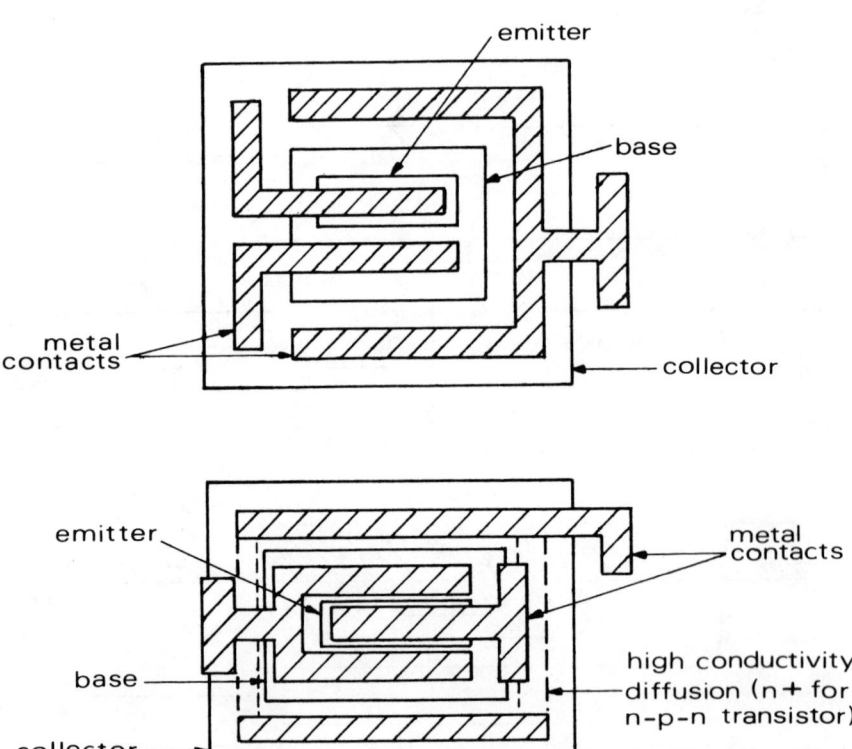

Figure 5.12 Low power high-frequency transistors.

When the epitaxial diffused process is combined with the buried-layer technique, transistors with collector resistances approaching that of discrete devices can be achieved. In the buried-layer technique, a low resistivity layer is formed in the collector region under the collector junction. This is done before the deposition of the epitaxial layer. The entire process is shown in Fig. 5.11 a through d.

Another technique for producing integrated transistors is the oxide isolation process. A thin layer of silicon dioxide, as shown in Fig. 5.12, replaces the *pn* junction around the collector region. In terms of transistor electrical characteristics, this technique is superior to the *pn* junction isolation technique, because parasitic collector-to-substrate capacitance is reduced substantially, and *pnp* parasitic action is eliminated. However, the main disadvantage of this technique is additional processing steps, which naturally increase the cost and reduce the yield of the device.

The preceding discussion has been carried out mainly with reference to *npn* transistors, because they are superior in performance to their counterpart *pnp* transistors. Due to the fact that electron mobility is larger than hole mobility, *npn* transistors have slightly higher $f_T$ values than do *pnp* devices of similar dimensions. However, *pnp* transistors are found in integrated circuits, and can be fabricated by interchanging all *p*- and *n*-type regions.

## 5.6 Design Considerations

The following points must be taken into account during the design and layout of an integrated transistor:

1. The emitter area and periphery should be large enough to provide the desired current.

2. The emitter width should be small enough to reduce emitter crowding. Emitter crowding is a considerable factor at higher frequencies, and contributes to a reduction in gain.

3. The base area should be only slightly larger than the emitter area, and the collector area only slightly larger than the base area. This design approach will reduce junction and isolation capacitance.

4. Emitter, base, and collector areas must be large enough to allow contact to their corresponding regions.

5. The average power generated per unit area must be low enough to be dissipated by the device.

The preceding rules may be interpreted into more general terms by saying that, for low-power, high-frequency transistors, small physical size is important. The layout in this case is determined mainly by how small an area can be masked that permits metalized contacts. The latter again is a factor determined by photoengraving techniques.

Maximum current capacity of a transistor is limited by factors such as emitter crowding. Due to the latter factor, the maximum current is normally proportional to the emitter periphery rather than the size of the transistor. The important requirement for practical transistors, particularly high-current transistors, requires a large ratio of emitter periphery to emitter area. Additional factors that must be taken into account during the layout are the maximum frequency of oscillation and the current level. The alpha cutoff frequency and the current-gain-bandwidth frequency are basically, but not completely, independent of the emitter area. There is a partial dependence of $f$ and $f_T$ on layout.

## 5.7 Integrated-Transistor Types

The majority of transistors used in integrated circuits are low-power, high-frequency devices. Two such types are shown in Fig. 5.13. The transistor in Fig. 5.13a has an emitter area of 1 by 2 mils. The overall size, including isolation diffusion, for such a transistor is about 10 mils$^2$. The other type of transistor, shown in Fig. 5.13b, operates at about twice the current level. For a maximum beta of 50 and a base resistivity of 0.2 $\Omega$-cm, current levels at maximum beta would be about 5 and 10 mA for the illustrated transistors, respectively.

The circuit contacts are made to the emitter, base, and collector regions by means of an alloyed vacuum evaporated aluminum thin film. The major factor in the design is that, during the alloying process, aluminum tends to dope silicon $p$ type. After alloying, the doping level is at the solid solubility of aluminum in silicon. This gives a doping level of about $5 \times 10^{18}$ atoms/cm$^3$, and, although it presents no difficulty for $p$-type material (it merely makes it more $p$ type), it creates a rectifying $pn$ junction for $n$-type material. The collector region of $npn$ transistors and the base region of $pnp$ transistors are typically doped lower than the $5 \times 10^{18}$ atoms/cm$^3$ level; to prevent this rectifying action, a shallow $n+$ enhancement diffusion is made in the lightly doped $n$-type regions prior to metalization. In $npn$ transistors, this is achieved by the same diffusion as used for the emitter; for $pnp$ transistors, an additional diffusion step is required.

Figures 5.14a and b illustrate a combination of two transistors

## Integrated-Transistor Types 55

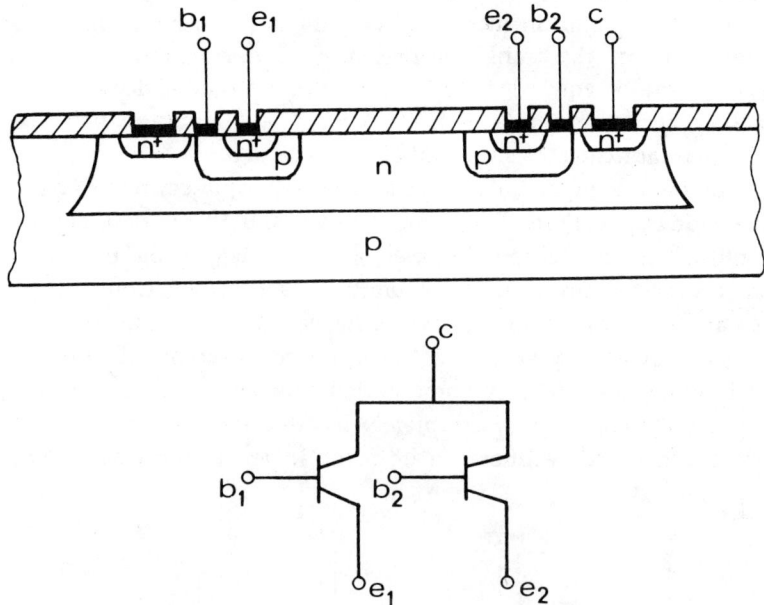

Figure 5.13 Multiple transistor structure.

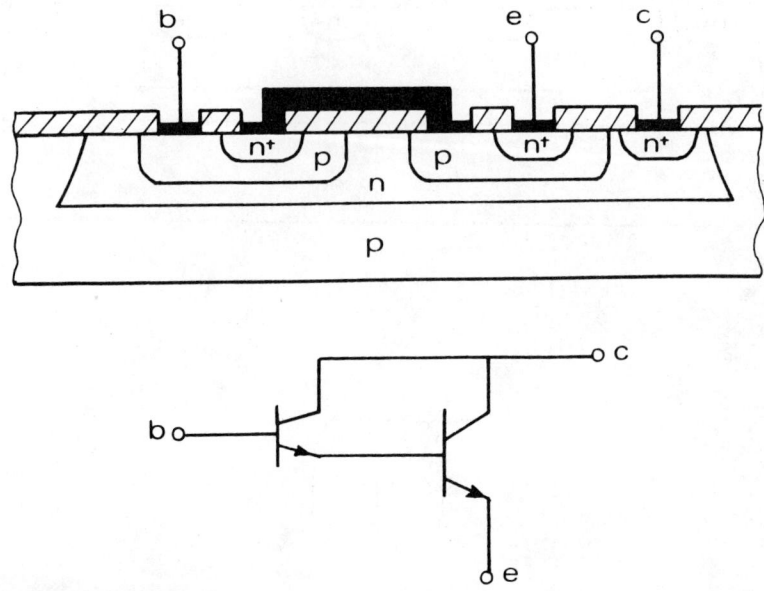

Figure 5.14 Darlington connection.

## 56 Integrated Diodes and Transistors

on the same substrate and their equivalent schematic. Such combinations are common in integrated circuits, and can be made by combining some of the transistor regions in a device and producing a structure that is equivalent to two or more discrete devices connected together. In the example of Section 5.2, the Darlington pair would have been fabricated as shown in Fig. 5.15a and b.

In some logic circuits, such as TTL, multiple emitter transistors have wide applications. For these transistors, both the base and emitter diffusions can be combined, as shown in Fig. 5.16a and b. With such a process, any number of emitters is possible. Combining the base and collector regions provides lower total isolation capacitance, simpler diffusion masks, and fewer interconnections. Figure 5.17a and b shows an *npn–pnp* combination that again offers advantages over two interconnected, completely isolated transistors.

A controlled rectifier can be made in an IC form as shown in

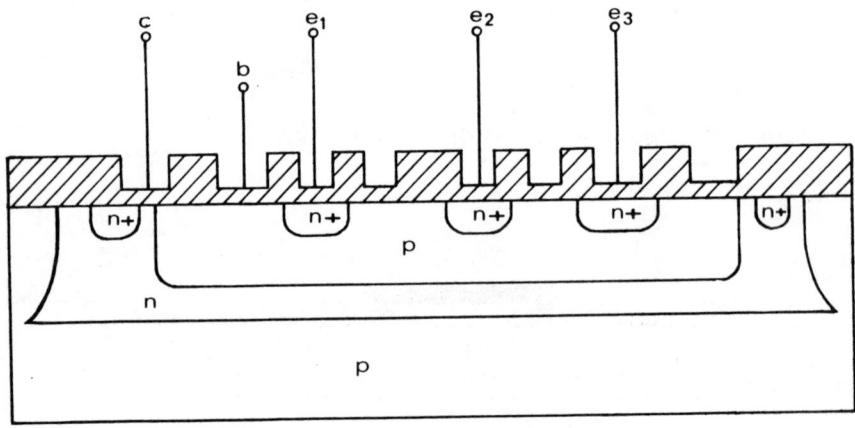

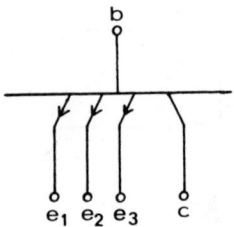

Figure 5.15 Multiple emitter transistor.

Integrated-Transistor Types  57

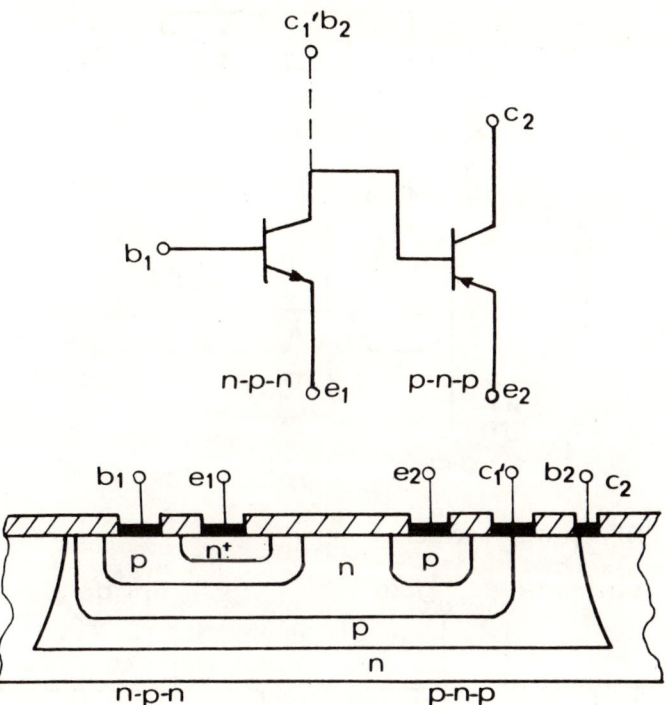

Figure 5.16  Direct-coupled n-p-n—p-n-p transistors.

Fig. 5.18. The *p–n–p–n* device (controlled rectifier) is a bistable unit consisting of four semiconductor regions and three *pn* junctions, with alphas of $\alpha_n$ and $\alpha_p$, respectively. Any four adjacent regions in an IC are capable of *p–n–p–n* switching action if they are close enough for significant minority carrier interaction between the junctions. The requirement for a low impedance state is $\alpha_n + \alpha_p = 1$, plus the correct voltage polarity. If one of the transistors, for example, has a high alpha value, the other transistor region can have a low alpha value and still provide a low impedance state.

When *p–n–p–n* switches are required in an IC, there are several ways of fabricating them, as shown in Fig. 5.19. The first switch is formed in a conventional integrated structure, and, owing to the fairly poor *pnp* characteristics of such a structure, this type has a high holding current and a high *on* resistance. The second type is a

## 58  Integrated Diodes and Transistors

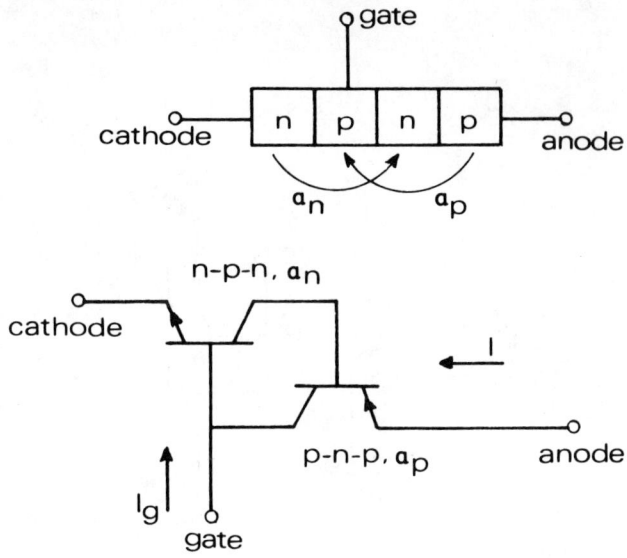

Figure 5.17  p-n-p-n switch.

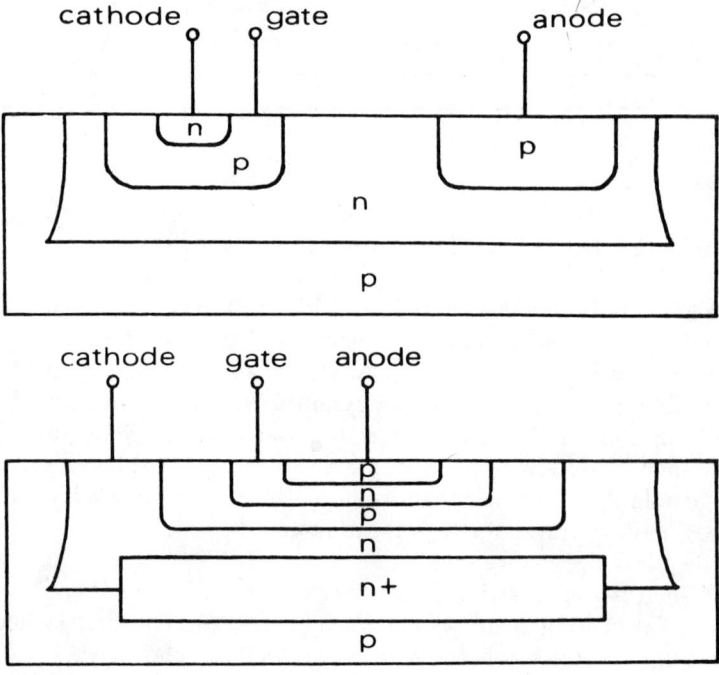

Figure 5.18  Two possible p-n-p-n structures for ICs.

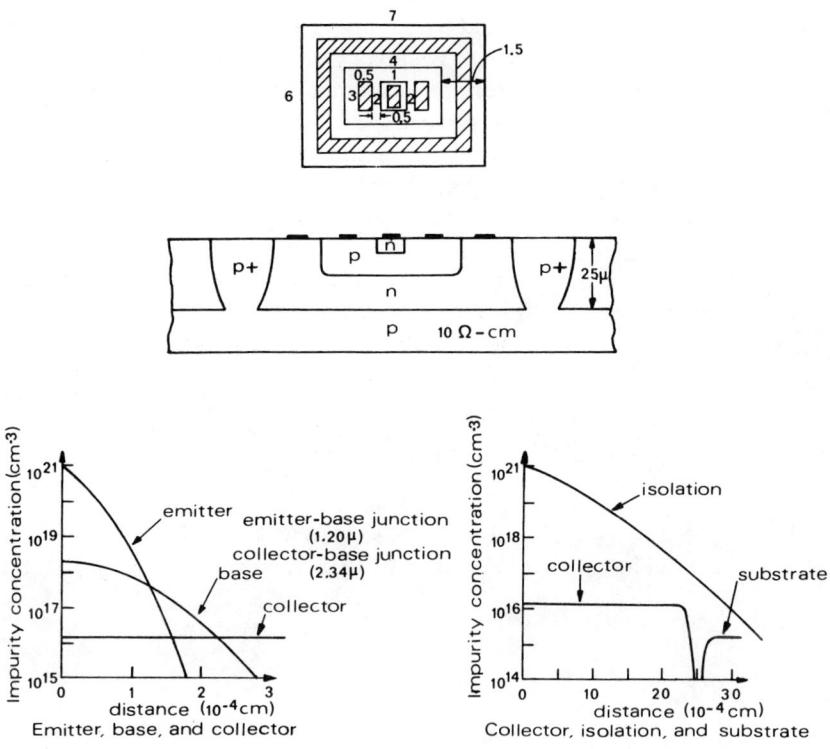

Figure 5.19 Low-current high-frequency example.

more conventional one, and does exhibit fair switching characteristics with a low *on* resistance.

## 5.8 Design Examples

The first example is a low-current, high-frequency transistor whose layout has been assumed to be as shown in Fig. 5.20. The transistor is a double-base contact transistor with a 1- by 2-mil emitter region, not including a buried layer. For this particular design, we assume that the maximum beta of the transistor is 50. The various parameters for evaluation are as follows:

$C_{TE}$ = total emitter capacitance
$C_{TC}$ = total collector capacitance
$C_{TS}$ = total substrate parasitic capacitance

$I_{C\beta max}$ = collector current at maximum beta
$f\alpha_b$ = grounded base cutoff frequency
$f_{max}$ = maximum frequency of oscillation
$r_{CS}$ = collector saturation resistance
$r'_b$ = base resistance
$BV_{CBO}$ = collector–base breakdown voltage
$BV_{EBO}$ = emitter–base breakdown voltage
$BV_{CSUB}$ = collector–substrate breakdown voltage

Let us first calculate the junction capacitances. At equilibrium, the emitter junction capacitance per unit area was found to be, say, $1.05 \times 10^5$ pF/cm², and the equilibrium voltage is, say, 0.89 V. This gives

$$\frac{C_{TE}}{A_e} = 1.05 \times 10^5 \text{ pF/cm}^2 \left(\frac{0.89}{0.89 \text{ V} - V_{BE}}\right)^{1/3}$$

The capacitance for the collector–base junction is a function of collector–base voltage. For $V_{BC} < -1$, this can be approximated by

$$\frac{C_{TC}}{A_c} = 8 \times 10^3 \text{ pF/cm}^2 \left(\frac{10 \text{ V}}{V_{CB}}\right)^{1/2}$$

The other capacitance to be evaluated is the isolation capacitance. The bottom part of the isolation junction approximates a step junction, and the capacitance $C_{SB}$ is given by

$$\frac{C_{SB}}{A_{sb}} = 3.47 \times 10^3 \text{ pF/cm}^2 \left(\frac{10 \text{ V}}{V_{CS}}\right)^{1/2}$$

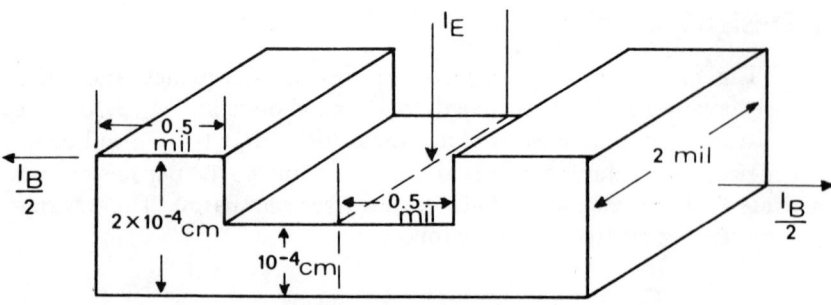

Figure 5.20 Base-collector resistance sketch.

where $A_{sb}$ is the bottom area of the isolation junction and $V_{CS}$ is the collector-to-substrate reverse bias. The capacitance has been normalized to the value at $V_{CS} = 10$ V. The sidewall component of the isolation junction has a different capacitance determined by the collector doping. This is evaluated as

$$\frac{C_{SW}}{A_{sw}} = 1.01 \times 10^4 \text{ pF/cm}^2 \left(\frac{10 \text{ V}}{V_{CS}}\right)^{\frac{1}{2}}$$

where $A_{sw}$ is the sidewall area.

The areas of the various $k$ junctions are evaluated as

$$A_e = 1.36 \times 10^{-5} \text{ cm}^2$$
$$A_c = 8.46 \times 10^{-5} \text{ cm}^2$$
$$A_{sb} = 2.71 \times 10^{-4} \text{ cm}^2$$
$$A_{sw} = 2.29 \times 10^{-4} \text{ cm}^2$$

The edges of the emitter and collector junctions are included in the area calculations. The junction capacitances are then evaluated as

$$C_{TE} = 1.43 \text{ pF} \left(\frac{0.89}{0.89 \text{ V} - V_{BE}}\right)^{\frac{1}{3}}$$

$$C_{TC} = 0.67 \text{ pF} \left(\frac{10 \text{ V}}{V_{CS}}\right)^{\frac{1}{2}}$$

$$C_{TS} = 3.27 \text{ pF} \left(\frac{10 \text{ V}}{V_{CS}}\right)^{\frac{1}{2}}$$

where $C_{TS}$ is the total isolation parasitic capacitance. For a forward bias of 0.7 V at the emitter, and $V_{CB} = V_{CS} = 10$ V, the capacitances are

$$C_{TE} = 2.37 \text{ pF}$$
$$C_{TC} = 0.67 \text{ pF}$$
$$C_{TS} = 3.27 \text{ pF}$$

The next step is calculation of the base and collector resistances. The sketch of Fig. 5.21 is used for this calculation. The resistance of the base is given by the base region and is evaluated at low current where the emitter current is uniform over the emitter area, as

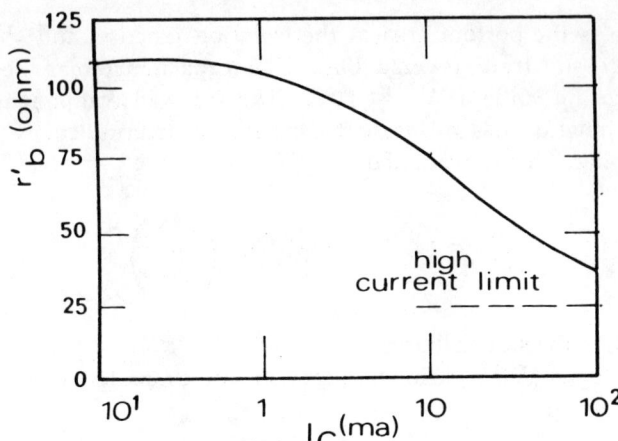

Figure 5.21  Theoretical variation of base resistance with collector current (assumed $B_{max} = 50$).

$$r'_b = \frac{1}{2}\left[\frac{1}{3}\frac{(0.2\ \Omega\text{-cm})(0.5\ \text{mil})}{(10^{-4}\ \text{cm})(2\ \text{mil})} + \frac{(0.04\ \Omega\text{-cm})(0.5\ \text{mil})}{(2\times 10^{-4}\ \text{cm})(2\ \text{mil})}\right]$$

$$= 108\ \Omega$$

where 0.2 $\Omega$-cm is the average resistivity of the base region under the emitter, and 0.04 $\Omega$-cm is the average resistivity of the base not under the emitter. At high currents, the emitter current is crowded near the emitter periphery, and $r'_b$ nears 25 $\Omega$. Using this analysis, we arrive at the diagram of Fig. 5.22.

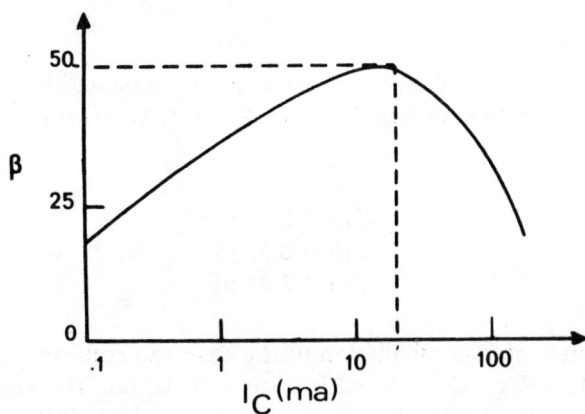

Figure 5.22  Expected variation of B with collector current.

The collector resistance is given by

$$r_{cs} = \frac{0.5 \text{ }\Omega\text{-cm}}{2 \,(25 \times 10^{-4} \text{ cm})} \frac{18 \text{ mil}}{6 \text{ mil}} = 35 \text{ }\Omega$$

where 0.5 $\Omega$-cm is the collector resistivity.

The collector current at which maximum beta occurs determines the useful current range of the transistor. Useful betas occur for currents about one order of magnitude above and two orders of magnitude below this value. Thus this current can be given by

$$I_{C\beta\text{max}} = \frac{0.26 \text{ V} \,(1.02 \times 10^{-2} \text{ cm})}{0.2 \text{ }\Omega\text{-cm}} (100)^{1/2} = 13 \text{ mA}$$

Based upon this value, and the assumed value of $\beta_{\text{max}} = 50$, Fig. 5.23 shows a sketch of how a beta would be expected to depend upon collector current. The useful current range of the transistor should then be from about 0.1 to about 100 mA.

The grounded cutoff frequency is given by

$$\frac{1}{f\alpha_b} = 2\,(\tau_e + \tau_b + \tau_x + \tau_c)$$

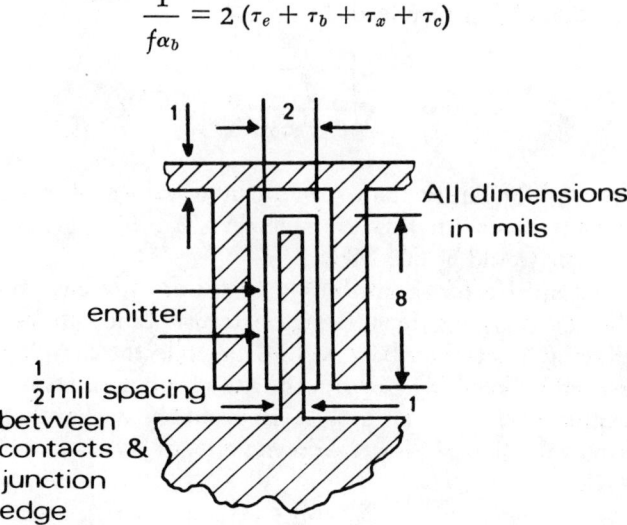

Figure 5.23  Emitter strips used in forming high current transistor.

where

$$\tau_e = \frac{kTC_{TE}}{q_2 I_E}$$

$$\tau_b = \frac{w_b}{2.43 D_{nb}}$$

$$\tau_x = \frac{x_m}{2 V_{sc}}$$

$$\tau_c = r_{CS}(C_{TC} + C_{TS})$$

The present analysis is evaluated at collector–base and collector–substrate voltages of 10 V and a collector current of 5 mA. The base width and collector depletion region width $x_m$ are $0.67 \times 10^{-4}$ cm and $10^{-4}$ cm.

In the base region, $D_{nb} = 13.3$ cm$^2$/sec. With these values, the preceding parameters are evaluated as

$$\tau_e = 0.12 \times 10^{-10} \text{ sec} \qquad \tau_x = 0.06 \times 10^{-10} \text{ sec}$$
$$\tau_b = 1.39 \times 10^{-10} \text{ sec} \qquad \tau_c = 1.44 \times 10^{-10} \text{ sec}$$

These combine to give

$$f\alpha_b = 530 \text{ MHz}$$

An evaluation of $f_{\max}$ then leads to

$$f_{\max} = \sqrt{\frac{f_T}{8\pi r_b' C_{TC}}} = 1.1 \text{ GHz}$$

which is slightly larger than experimental values of $f_{\max}$ for comparable dimensions (in this example, $W_E/2 + S' = 1$ mil). A more realistic value would be 800 MHz.

The major factors contributing to $f\alpha_b$ are the base transit time and collector charging time. The base transit time can be decreased by employing a narrower base width; the collector charging time can be decreased by lowering $r_{CS}$ with the use of a buried layer. The theoretical value of $f_T$ as well as $f_{\max}$ is probably a slightly optimistic value. An evaluation of the breakdown voltages gives

$$BV_{EBO} = 5 \text{ V}$$
$$BV_{CBO} = 50 \text{ V}$$
$$BV_{CSUB} = 50 \text{ V}$$

The next example is a high-current transistor. It is assumed that the transistor should have a current level of 500 mA at $\beta_{max}$. Along with the current rating, other parameters that are assumed to be specified are

$$I_{C\beta max} = 500 \text{ mA} \qquad \beta_{max} = 50$$
$$f\alpha_b = 500 \text{ MHz} \qquad BV_{CBO} \geqq 25 \text{ V}$$
$$\qquad\qquad\qquad\qquad BV_{CSUB} = 25 \text{ V}$$

From the desired value of $f\alpha_b = 500$ MHz, it is seen that the transistor is to be a fairly high frequency transistor.

To simplify the design approach, we shall assume that the parameters received for the lower-power transistor are accurate, and some of them will be used in the second example. Thus a value of 530 MHz was received for $f\alpha_b$. Since this frequency is independent of emitter area, the same junction depths and impurity concentrations should result in an acceptable value of $f\alpha_b$ for the high-current transistor. To increase the cutoff frequency slightly and provide an added margin of assurance, a buried layer under the collector will be used in this design. Another result that we can use from the previous example is the impurity diffusions of the low-power transistor, which also give acceptable values of breakdown voltage and the desired value of maximum beta.

Thus, using the information from the low-power transistor, the design becomes a matter of determining the layout that provides the desired current level and, finally, verifying that the transistor does in fact give an $f\alpha_b$ of 500 MHz. The emitter or collector current can be given by

$$I_E/h = \frac{kT}{9_b} (2\beta_{max})^{\frac{1}{2}}$$
$$= 1.30 \text{ A/cm} \quad (\text{at } \beta_{max})$$

The total emitter periphery required to give 500-mA current is thus 0.385 cm. This will be divided into several emitter stripes. A typical emitter stripe is shown in Fig. 5.24. The width is shown to be 2 mils rather than the 1 mil used in the lower-power transistor, and the reason for this is to increase the reproducibility and ease in fabrication. The maximum length of the emitter stripe is determined by the voltage drop along the metalized emitter stripe as current flows to the emitter. To account for this effect, the sheet resistance of the deposited aluminum contact stripes must be known. The sheet resis-

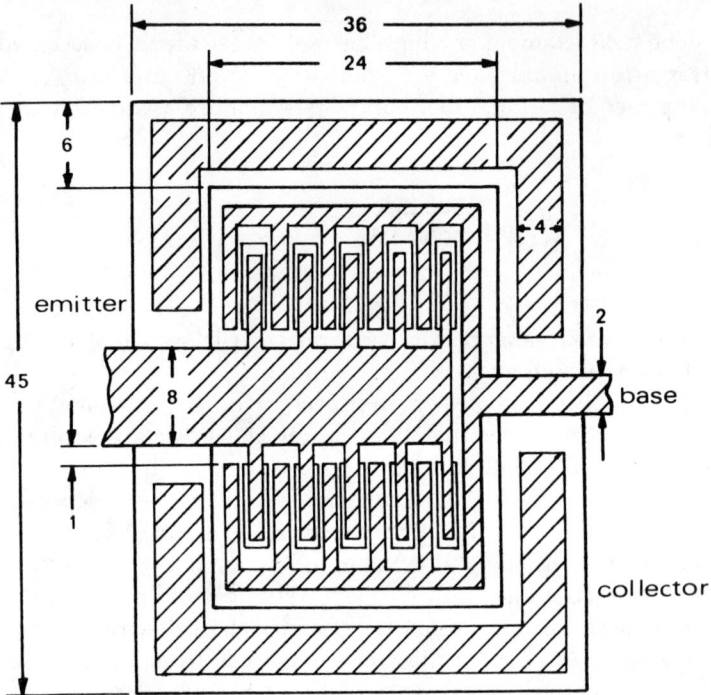

Figure 5.24  High current transistor.

tance of a 1-μm-thick aluminum deposit is assumed to be 0.028 Ω/square. For the present analysis, a sheet resistance of 0.05 Ω/square is used, corresponding to an aluminum deposit of slightly more than 0.5 μm in thickness. The current for a single emitter stripe, $I_l$, and the stripe length, $l$, must satisfy the inequality

$$I_l\,(0.05\ \Omega)\left(\frac{l}{1\ \text{mil}}\right) < 0.025\ \text{V}$$

This keeps the voltage drop along the emitter stripe less than 0.025 V. The emitter stripe current is also related to the stripe length by the current crowding expression

$$\frac{I}{2} = 1.30\ \text{A/cm}$$

Combining these equations, we get

$$l < 2.24 \times 10^{-2}\ \text{cm} = 8.8\ \text{mils}$$

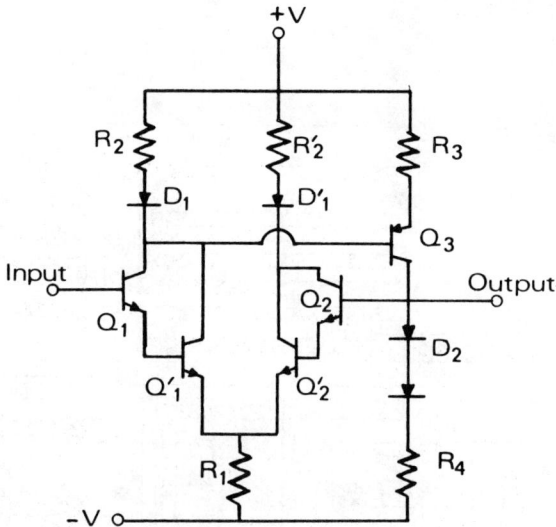

Figure 5.25 Schematic of buffer amplifier.

An emitter stripe of 8 mils, as shown in Fig. 5.23, is chosen for all emitter stripes. The minimum number of emitter stripes required for the particular transistor is thus $0.385 \text{ cm}/2l = 9.5$. A total of 10 emitter stripes 8 mils long will provide the desired current. The complete layout for such a transistor is shown in Fig. 5.25.

Figure 5.26 illustrates a buffer amplifier, and its integrated form is shown in Fig. 5.27.

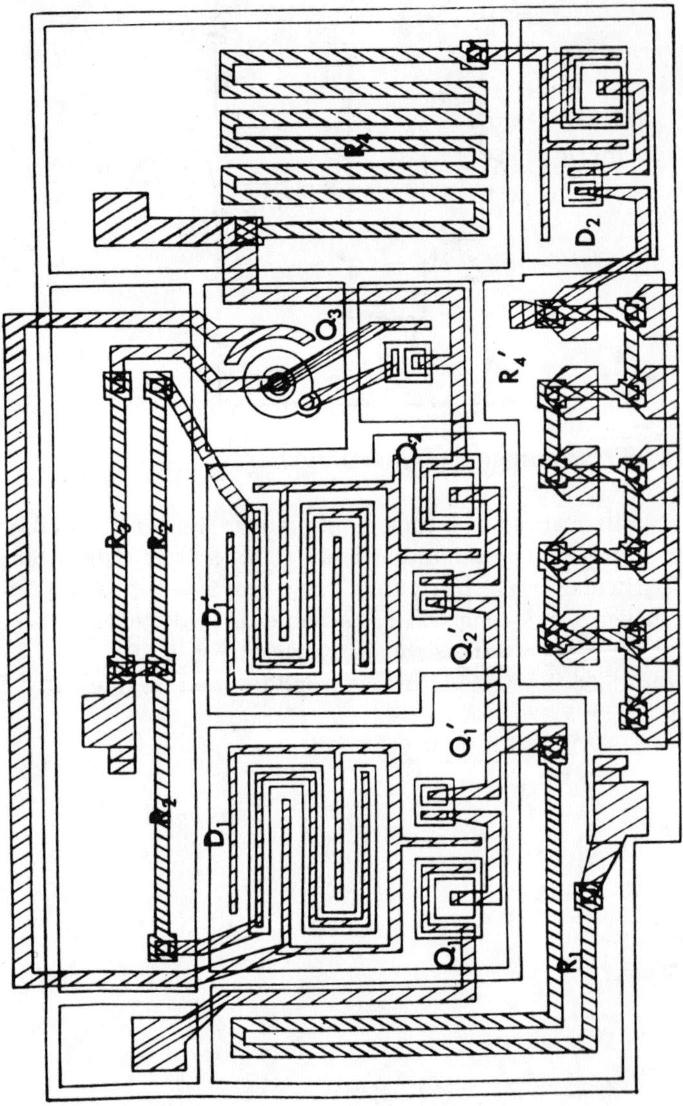

Figure 5.26

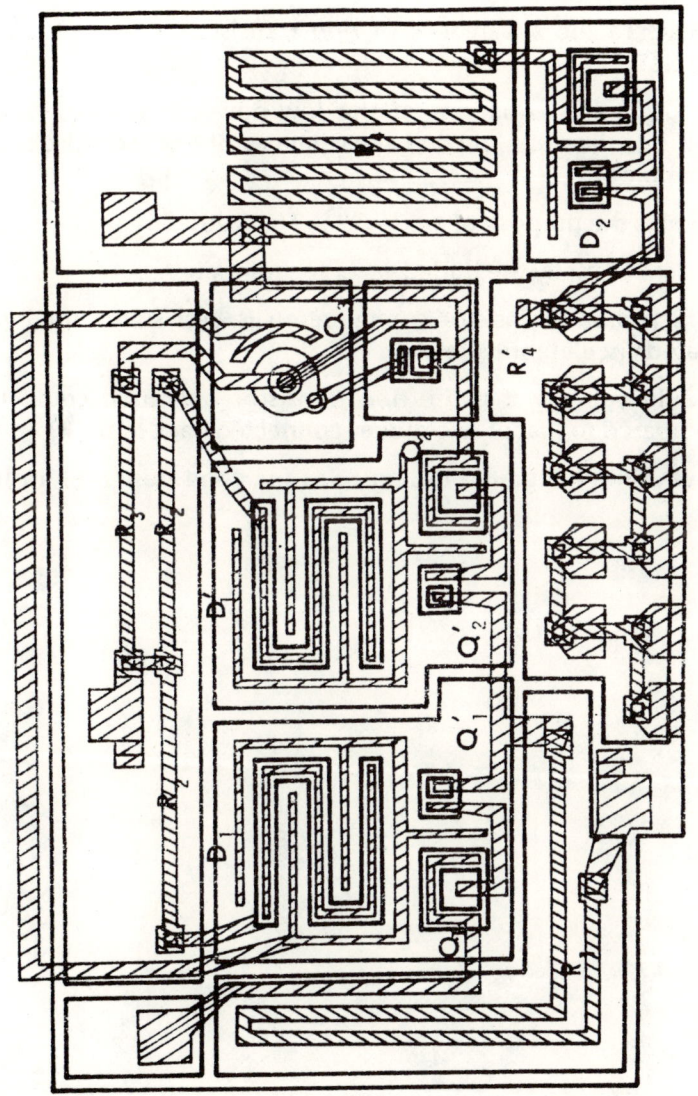

Figure 5.27

## Exercises

**5.1** What are the disadvantages of emitter crowding?

**5.2** What are the advantages of *npn* transistors over *pnp* transistors?

**5.3** By using the equations of the low-power transistor, could you calculate the junction capacitances of the second design example?

**5.4** What is the purpose of diode *D*2 in Fig. 5.25?

**5.5** Describe briefly the buried-layer technique.

**5.6** Which diode-transistor configuration is best suited for integrated circuit fabrication?

**5.7** Would you say that the *npn* transistor operation could be compared to that of two diodes connected back to back?

**5.8** Give some integrated design rules for good design of an IC.

# chapter six
# Metallic-Oxide-Semiconductor Circuits

## 6.1 General

We purposely did not include MOS transistors in the bipolar transistor section because they belong to a world of their own. Their operation differs substantially from that of bipolar devices. The MOS transistor is a *unipolar* device, since only one type of carrier is used in the operation of a particular transistor. For $p$-channel MOS, the carriers are holes, whereas electrons are the carriers for $n$-channel MOS.

Another difference is that bipolar transistors are bulk devices; that is, their active region lies in the base, several micrometers beneath the surface between the emitter and collector. On the contrary, MOS devices are surface-effect devices. Their active region consists of a channel that is induced (for enhancement-mode operation) at the silicon–silicon dioxide interface. MOS circuits have recently found extensive use in microprocessors, memories, and other circuits.

## 6.2 Basic MOS Transistor

The basic silicon-gate MOS transistor appears in cross section along with conventional circuit symbols in Fig. 6.1. There are two modes of operation for field-effect devices: enhancement and depletion. The devices shown are enhancement-mode, which are nonconducting with zero volts applied to the gate. A conducting channel is created (enhanced) by an electrostatic field. The field arises from the voltage applied between the gate and source terminals. For the $p$-channel device (Fig. 6.1a), negative voltages (with respect to source) are applied to the gate and establish the conducting channel. The channel develops between the source and drain, and at the interface between the gate dielectric and the substrate. For $n$-channel devices (Fig. 6.1b), conduction requires a positive gate-to-source voltage.

Depletion-mode devices normally conduct even with zero volts applied to the gate. These devices require the application of an appropriate gate-to-source voltage to deplete the channel, or turn them off. Most MOS circuit designs use enhancement-mode devices.

MOS transistors are field-effect devices because the gate is electrically isolated from any other part of the transistor. As a result, the dc input impedance (using the gate as the input) is extremely high, on the order of $10^{14}$ Ω. This high dc impedance characteristic primarily determines the nature of the MOS input surface.

Figure 6.2a shows the $V$–$I$ behavior of MOS enhancement-mode transistors. When the device operates in the region labeled nonsaturated, its $V$–$I$ characteristic is approximately resistive. The term *nonsaturated* means that the device is not conducting as much drain current as possible for a given gate-to-source voltage.

The device behaves like a current source or sink when operated in the saturated region of the curves. In this area, the drain current is no longer a function of the drain-to-source voltage. Note that the larger the gate-to-source voltage, the larger the saturation current becomes.

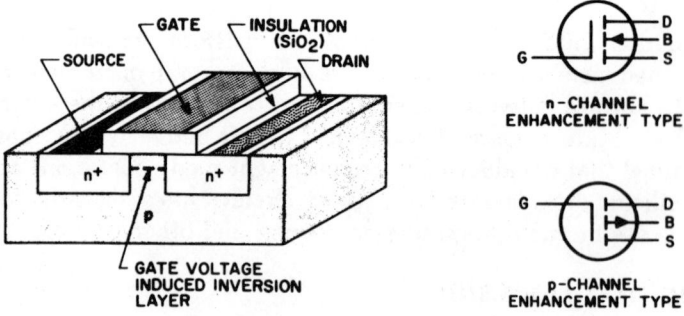

Figure 6.1 MOS transistor.

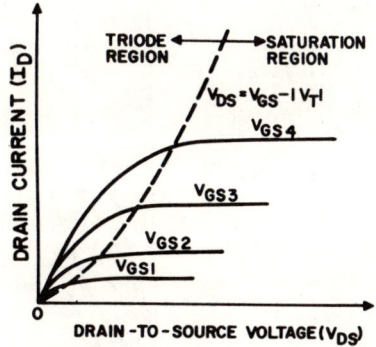

Figure 6.2 Drain-to-source current as a function of drain-to-source voltage.

The input-transfer curves (Fig. 6.2b) show the drain current as a function of gate-to-source voltage. The device is off ($I_D = 0$) until $V_{GS} \geq V_T$. Low-threshold MOS processes, such as silicon-gate, allow MOS logic to be compatible with bipolar families. These families include standard TTL and other common bipolar logic lines with switching thresholds of about 1.5 V.

For $p$-channel silicon-gate devices, $V_T$ is nominally 2 V; $n$-channel silicon-gate devices have a $V_T$ of 1 V. The threshold voltage has a typical variation with junction temperature of about $-2$ mV/°C from $-50$ to 125°C. Clearly, the threshold voltage is not very sensitive to temperature variations.

The slope of the input transfer curves is the transconductance, $g_m$. This parameter can be used as the gain figure or transfer function for the MOS transistor. It governs the output current of a device for a given input voltage. Transconductance decreases with rising junction temperature. Current gain (beta) in bipolar devices increases with rising temperature.

## 6.3 MOS Applications

As previously stated, MOS transistors have found wide applications in logic and memory circuits. As an example, observe Fig. 6.3, which shows how $p$-channel MOS transistors operate in the NOT logic function. The upper transistor in each inverter functions totally as a load resistor and uses the transistor's transconductance to get a large-value resistor in the relatively small area required by the device. Otherwise, resistors composed of fairly low resistivity materials would be needed on the chip, and resistors tend to occupy

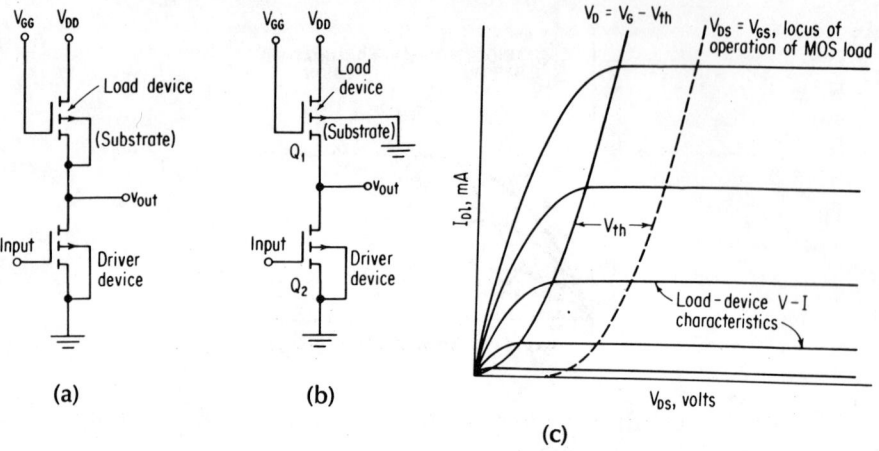

Figure 6.3  MOS inverters and load-device characteristics.

quite a lot of space. High resistances are often required to minimize the power consumed in complex circuits. A high load resistance can be achieved by the use of MOS transistors because the gate voltage governs the channel resistivity. Incidentally, the difference in the two preceding circuits is in the mode of operation of their load devices. The resistor in Fig. 6.3a is in saturation and thus behaves like a current source; the load resistor in Fig. 6.3b is nonsaturated and behaves like a resistor. Due to their high input impedance, MOS devices are ideally suited for use in dynamic memory-cell designs. In this type of

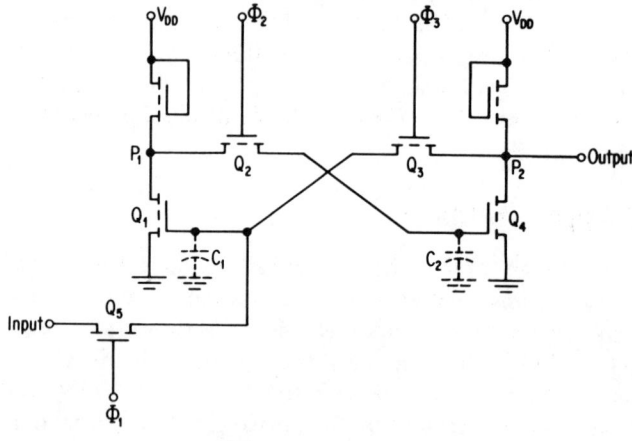

Figure 6.4  MOS R-S Flip Flop.

cell, a parasitic capacitance stores charge that is periodically refreshed. The refresh compensates for the small leakage current at the storage nodes. By comparison, a static memory cell continuously refreshes the charge through load resistors.

A simple way to obtain a memory function is to use the flip flop. Figure 6.4 shows the MOS version of a basic *R–S* flip flop. The circuit uses *p*-channel MOS devices, and consists of two cross-coupled inverters. Transistors Q1 and Q4 provide access to the memory cell. The *on* resistance ratio of Q1 and either Q2 or Q3, depending on the cell state, determine the logic 1, or most-positive-voltage, level. The logic 0 is $-V + V_T$.

## Exercises

**6.1** What is the difference between bipolar and MOS structures?

**6.2** Draw the equivalent bipolar circuit for Fig. 6.4.

**6.3** Draw the equivalent circuit for Fig. 6.3.

**6.4** What is depletion mode and what is enhancement mode? What are the disadvantages or advantages of either type?

**6.5** Do you see any differences between the processing of bipolar devices and MOS devices by studying Fig. 6.1?

**6.6** Describe an important parameter of MOS devices.

chapter seven
# Basic Design Rules and Equations

## 7.1 Introduction

The theory discussed so far has provided a rather thorough study of the design and fabrication of the different integrated competents. The discussion has been written, as much as possible, free of complicated equations. However, there are certain design rules and mathematical equations that are very essential in the design and fabrication of IC's. It was thought best that these rules and equations be collected in one chapter rather than scattered throughout the text. And this precisely is the purpose of this chapter.

## 7.2 Mathematical Equations

### Lifetime and Diffusion Length

One of the properties of silicon is defined by the time a minority carrier will exist before it recombines with a majority carrier. This property is called *lifetime*, and can be best given in terms of the *diffusion length*.

## Basic Design Rules and Equations

Thus

$$L_D^{-1} = (D\tau)^{-1/2}$$

where

$L_D$ = diffusion length
$D$ = diffusion constant ($\mu kT/q$)
$\tau$ = lifetime (sec)

## Junction Capacitance

A depletion layer may also be considered as the spacing between two charged parallel plates, and, in that case, the capacitance per unit is

$$C = \sqrt{\frac{qk\varepsilon_0 N_D}{2V}}$$

The equation shows that the capacitance of a *pn* junction is a function of the applied voltage. Thus the *pn* junction is sometimes used as a capacitive element in a circuit, but will be polarized, since the presence of the depletion region depends on reverse biasing the junction, and its characteristics will be voltage dependent.

## Diode Equation

As described in previous chapters, when a dc bias voltage is applied to a *pn* junction, both hole ($I_p$) and electron ($I_n$) currents are created, and, naturally, the total current ($I$) is the sum of these two currents. Thus

$$I = I_p + I_n$$

The basic voltage–current relation for the *pn* junction is

$$I = \left(\frac{qAD_p p_n}{L_p} + \frac{qAD_n n_p}{L_n}\right)(e^{qV/kT} - 1)$$

where $A$ is the area of the junction in square centimeters. Normally, one side of the junction is doped more lightly than the other, and therefore the term associated with that side is numerically insignificant as compared to the other. Thus the preceding equation may be reduced as follows:

$$I = \left(\frac{qAD_p p_n}{L_p}\right)(e^{qV/kT} - 1)$$

## Reverse Current, Forward Voltage, Breakdown Voltage

The reverse current equation for a *pn* junction is

$$I_R = \frac{qAD_p p_n}{L_p}$$

where $A$ is the junction area, $D_p$ is the diffusion constant of holes diffusing in *n*-type material, and $L_p$ is the diffusion length or the mean distance a minority carrier will travel before recombination. $D_p + L_p$ and $p_n$ have relationships fixed by the doping level. The forward-voltage equation is

$$V_F = \frac{kT}{q}\left(\ln \frac{I_F}{I_R}\right) = 26 \ln \frac{I_F}{I_R} \text{ (mV)}$$

where $\frac{kT}{q}$ is about 25 mV at room temperature. The equation indicates that forward voltage is a direct function of temperature.

Finally, the breakdown voltage, taking into consideration the doping levels that exist on either side of the junction, is

$$V_B = \frac{k\varepsilon_0}{2qN_D}(\varepsilon_{max})^2$$

where
 $k$ = dielectric constant (12 for silicon)
 $\varepsilon_0$ = permittivity of free space ($8.85 \times 10^{-14}$ F/cm)
 $q$ = electron charge ($1.6 \times 10^{-19}$ C)
 $\varepsilon_{max}$ = field strength necessary to cause ionization
 $N_D$ = density of electron donor atom on more lightly doped or high resistive side

## Emitter Efficiency

The emitter efficiency of the transistor is given by

$$\gamma = \frac{1}{1 + \frac{\rho_e W}{\rho_b L_{ne}}}$$

where
 $W$ = effective base width (cm)
 $\rho_e$ = emitter resistivity ($\Omega$-cm)

$\rho_b$ = base resistivity (Ω-cm)
$L_{ne}$ = electron diffusion length in the emitter (cm)

For most conditions, it can be shown that

$$\frac{\rho_e}{L_{ne}} \approx R_{EE}, \text{ the emitter sheet resistance}$$

and

$$\frac{\rho_b}{W} \approx R_{BB}, \text{ the base sheet resistance}$$

thus

$$\gamma = \frac{1}{1 + (R_{EE}/R_{BB})} = \frac{1}{1 + \frac{2.5}{200}} = 0.988$$

## Emitter Resistance and Transition Capacitance

The emitter resistance is given by

$$r_e = \frac{kT}{qI_E} = \frac{25}{1 \text{ mA}} = 25 \text{ Ω}$$

The capacitance is given by

$$C_{Te} = A_e \sqrt{\frac{qk\varepsilon_0 N'_B}{2V}}$$

where

$A_e$ = emitter area (cm²)
$N'_B$ = impurity concentration of base under the emitter
$V$ = voltage across the junction

## Base Transport Factor

The base transport factor ($\beta^*$) is given by

$$\beta^* = 1 - \frac{1}{2}\left(\frac{W}{L_{nb}}\right)^2$$

where $L_{nb}$ is the electron diffusion length in the base and $W \ll L_{nb}$. This $\beta^*$ should not be confused with the transistor current gain (beta without the asterisk). Certain of the above equations have been already met in the text, but they can be more easily memorized in this chapter.

## 7.3 Practical Designs

Fabrication of the various integrated components has been discussed in the previous chapters, and a summarized practical design is provided next.

The major fabrication steps of integrated circuits are as follows:

**(a) Preparation of p-Type Substrate and Substrate Surface.** This initial fabrication step starts with the growth of an epitaxial $n$-type layer on a $p$-type silicon surface, also providing part of the isolation function.

**(b) Isolation Diffusion**  Isolated regions are formed by diffusing $p$-type channels through the epitaxial layer. A silicon dioxide layer is thermally grown over the epi layer, and, by photoengraving, holes are etched through the $SiO_2$ wherever an isolation channel is desired. Next the wafer is placed in a diffusion oven, and, at a temperature of about 1300°C, sufficient impurities diffuse into the wafer, and invert the $n$-type layer to $p$-type. By maintaining the substrate at a negative potential with respect to all $n$-type regions, these regions are electrically isolated from each other as well as from the substrate, with the exception of the reverse leakage current and capacitance of the isolation junction. During the isolation diffusion, the wafer again is totally covered by $SiO_2$.

**(c) Base Diffusion**  The transistor base diffusion is formed by first etching holes in the silicon dioxide layer over all regions that are to be $p$-type bases. Then $p$-type impurities are diffused through these holes, forming the collector–base junction.

**(d) Emitter Diffusion**  The emitter diffusion is formed in a manner similar to the base diffusion, with the exception that, in this case, the impurities are electron donors. During the high concentration emitter diffusion, an $n+$ region is also formed in the $n$-type collector region so that ohmic contact can be made to the aluminum. Otherwise, the aluminum atoms (electron acceptors) that diffuse slightly into the surface of the silicon can invert high-resistivity $n$-type silicon, and result in $pn$ junctions at collector contacts. Next, holes are etched through the dioxide layer at all points where contact is to be made to collector, base, and emitter.

**(e) Aluminum Metallization**  Next aluminum is deposited over the entire surface of the wafer, and is selectively etched away, leaving the desired interconnection pattern.

## 7.4 Practical Design of Resistors

The most important factor to consider during the design of integrated resistors is their fabrication compatibility with that of the other components in the circuit. Diffused resistors are naturally a logical choice, mainly owing to their similarity in fabrication with that of transistors. However, there are cases where space is a problem, and thus thin-film resistors are considered owing to their higher sheet resistivity over diffused resistors (provided area and resistor width are held constant). The same resistors are considered where improved tolerances, high-frequency response, and lower temperature coefficients are required. Design techniques, as well as the advantages and disadvantages of both these types are discussed next.

Resistance of a typical resistor is calculated by the sheet resistivity ($\rho_s$) and the ratio $l/w$. The electrical contact between the resistor and other points in the circuit is accomplished by etching holes in the dioxide layer and depositing aluminum interconnections.

The following guidelines are helpful in designing integrated diffused resistors.

1. Base or emitter diffusion.

    (a) $R < 20\ \Omega$, emitter diffusion recommended.
    (b) $20 < R < 30\ \text{k}\Omega$, base diffusion recommended.
    (c) $30\ \text{k}\Omega < R$, FET use recommended.

2. Geometry.

    (a) Minimum width ($w$) equals 1 mil.
    (b) Length-to-width ratio should be larger than 10.
    (c) Power dissipation per square mil should be uniform for all the resistors in the circuit. Power dissipation can be obtained by the following equation:

    $$\frac{\text{area of } R_n}{A_{R_t}} = \frac{I^2_{rmsn} R_n}{\Sigma I^2_{rmsi} R_i}$$

where

$R_n$ = $n$th resistor in the circuit
$I_{rmsn}$ = rms current through $R_n$
$A_{R_t}$ = total area of all resistors in the circuit

   (d) Minimum area of aluminum contact with diffused regions is 1 mil$^2$.

(e) Diffused region should extend 0.5 mil beyond the aluminum contact area.
(f) All bends and turns should be shorted by aluminum metallization.

3. Resistor ratios.

    (a) Same diffusion recommended.
    (b) Locate near each other on substrate.
    (c) Resistance values should be approximately the same.

4. Precision resistors.

    (a) As large as possible.
    (b) Low-resistance segments should be added for adjustment, if practical.

5. Resistance calculations.

    (a) If $l/X \gg 1$, $R = \rho_s\, l/w$. As $l/X$ decreases, errors become significant in this equation. For a 2 μm $X$, the length of a resistor 1 mil wide with $\rho_s$ approximately 150 Ω/square should be larger than 5 mils in order to safely use this equation. At any rate, $l/X$ should be greater than 20.
    (b) If $l/w < 10$, the following equation is used for end effects:

$$R = \frac{\rho_s[l + 2(0.65)]}{w} \quad (l,w \text{ in mils})$$

$$R = \frac{\rho_s[l + 2(0.14)]}{w} \quad (l,w \text{ in mils})$$

Typical values of parasitic capacitance for these designs are 0.05–0.2 pF/mil² with leakage current of 10 nA/mil².

## Thin-Film Resistors

The resistance of a thin-film resistor is the product of sheet resistivity and the length-to-width ratio. The minimum width of this type of resistor should be about 1 mil, and tolerances to be expected are about ±10 percent. The power dissipation of such resistors should not exceed 3 mW/mil².

Table 7.1 provides information on integrated-circuit resistors.

## TABLE 7.1

| PARAMETER | DIFFUSED p | DIFFUSED n+ | NICHROME |
|---|---|---|---|
| Sheet resistance ($\Omega$/square) | 100–300 | 2.5 | 40–400 |
| Resistance per substrate area, 1-mil stripe, 1-mil spacing ($\Omega$/mil$^2$) | 50–150 | 1.25 | 20–200 |
| Temperature coefficient (ppm/°C) | +2800–1500 | 200–300 | ±100, ±10% |
| Power dissipation per active resistor area, depending on package and heat sink (mW/mil$^2$) | 3 | 3 | 2 |
| Maximum voltage (V) | 20 | 6 | — |
| Tolerance for high yield (%) | ±20 | — | ±8 |
| Distributed capacitance (pF/mil$^2$); negligible when using dielectric isolation | 0.2 | 0.6 | — |

## 7.5 Practical Design of Integrated Capacitors

There are several types of integrated capacitors. However, prior to finalizing a decision as to the use of a capacitor in an IC, one should ask whether or not that capacitor is really needed in the circuit. The designer should remember that capacitors are costly and occupy large areas on the wafer. If the circuit can be changed to avoid the capacitor, it should be.

The types of capacitors in IC's are the *pn* junction, the MOS capacitor, and the thin-film capacitor. As the diffused resistor, the junction capacitor is the logical choice, since its fabrication is compatible with that of transistors. However, the disadvantages of junction capacitors are that they are characterized by nonlinearities that can present problems in some applications; they have polarity, thus requiring a bias voltage; and, finally, the isolation capacitance associated with them may be substantial compared to that of the capacitor itself.

MOS capacitors also offer fabrication compatible to that of integrated circuits. Their capacitance per unit area is comparable to that of junction capacitors; but MOS capacitors are nonpolar, and the advantage to the latter is that the voltage can be set in such a way that any capacitance value may be derived, from a practically constant value to a highly nonlinear value.

Thin-film capacitors, although they require additional fabrication steps, offer the most significantly greater capacitance values by construction of multilayer capacitors. They are also nonpolar, offer good dissipation, parasitic and frequency characteristics, and are voltage independent.

The four different designs available for junction capacitors are as follows:

(a) Collector–base junction.

(b) Emitter–base junction.

(c) Collector–substrate.

(d) MOS structure on surface of circuit.

A rather important equation used in the calculation of capacitance is the following:

(1) Step-junction:
$$C/A = 2.93 \times 10^{-4} \left(\frac{N_\rho}{V} \text{ volt-cm}^3\right)^{1/2} \text{ pF/cm}^2$$

(2) Linear-graded junction:
$$C/A = 2.47 \times 10^{-3} \left(\frac{a}{V} \text{ volt-cm}^4\right)^{1/3} \text{ pF/cm}^2$$

where
$a$ = grade constant
$V$ = total junction voltage (including built-in potential)
$A$ = junction area
$N = N_\rho^{-1} = N_a^{-1} + N_d^{-1}$  ($N_a$ and $N_d$ are net concentrations of acceptors and donors on opposite sides of junction, respectively)

Since the junction capacitor is polarized, its junction must be reverse biased at all times, not exceeding the reverse breakdown voltage. Thus, as the capacitance per unit area increases, the working voltage range of the capacitor is decreased. Another factor related to the applied voltage is the effect the latter has upon the per unit area capacitance. The preceding equations indicate that capacitance varies with $V^{-1/2}$ and $V^{-1/3}$. Therefore, if signal voltages appear across the capacitor, it is important that the capacitance variation with voltage be minimized; that is, when distortion of a signal is present, a linear-graded (collector–base) junction should be used.

One of the optimum designs for an integrated capacitor is increase of its $C/A$ ratio. Figure 7.5 illustrates a parallel combination of the emitter–base and collector–base junctions that effectively increases the ratio while the breakdown voltage is not exceeded.

Two more factors should also be taken into consideration; they are the total junction area for optimum capacitance and the shape of the capacitor. In the first case, a maximum practical value is 500 mil$^2$ (about 500 pF). In the second case, the design should be such that the minimum area in the circuit should be occupied. Two desirable shapes for an integrated capacitor are the circular and the square shape. The circular shape has several advantages over the square in that it minimizes propagation distances in the circuit and improves the $Q$ of the capacitor.

MOS capacitors are easier to fabricate. Such a capacitor and its equivalent circuit are shown in Fig. 7.6. The capacitance per unit area is given by

$$C/A = \frac{0.33}{w_{ox}} \text{ pF/cm}^2$$

where $w_{ox}$ is the oxide thickness (cm). Table 7.2 lists the various characteristics for integrated capacitors.

## 7.6 Practical Design of Transistors

A typical integrated transistor is shown in Fig. 7.1. The several important factors to consider during design are signal gain, frequency response, and power-handling capacity; naturally, these factors are related to alpha, beta, resistances and capacitances associated with the collector, base, and emitter, and limiting quantities such as breakdown voltages, maximum currents, and temperatures.

Two typical transistors, their characteristics, and their geometries are shown in Figs. 7.2 and 7.3, respectively. The transistor surface geometry greatly influences the important factors given. Furthermore, size and shape can greatly simplify the design of interconnection patterns.

The optimum design of low-power transistors is to obtain maximum transistor current gain at low collector currents. This can be achieved when the transistor operates at collector currents at or near the point of maximum gain.

## TABLE 7.2
## INTEGRATED CAPACITOR CHARACTERISTICS

| PARAMETER | MONOLITHIC CAPACITORS | | | THIN-FILM CAPACITORS | | |
|---|---|---|---|---|---|---|
| | SINGLE-DIFFUSED $pn$ JUNCTION | DOUBLE-DIFFUSED $pn$ JUNCTION | THERMALLY GROWN $SiO_2$ | SiO | SILICATE GLASS | TANTALUM |
| pF/mil$^2$ | 0.1 when $V_{bias} = 0$ | 1.0 when $V_{bias} = 0$ | 0.25 | 0.01 | 0.4 | 2.5 |
| $V_{max}$ (V) | 30 | 6 | 50 | 50 | 50 | 20 |
| Dissipation factor (%) | | | | | | |
| 1 kHz | 10 | 100 | | 2.5 | 0.2 | |
| 1 MHz | | | 0.7 | 0.7 | 0.2 | 0.8 |
| 10 MHz | | | 2.0 | | 1.0 | 0.3 |
| Temperature co-efficient (ppm/°C) | Low | Low | Low | ±200 ±50 | +115 | +400 |
| Voltage sensitivity | $V^{-1/3}$ | $V^{-1/2}$ | 0 | 0 | 0 | 0 |
| Polar | Yes | Yes | No | No | No | No |
| Shunt capacitance, % component value | 25 | 25 | 18 | 0 | 0 | 0 |
| Leakage current at 5 V/A/pF | $10^{-9}$ | $10^{-9}$ | $10^{-9}$ | $10^{-14}$ | $10^{-15}$ | $10^{-15}$ |

## 88  Basic Design Rules and Equations

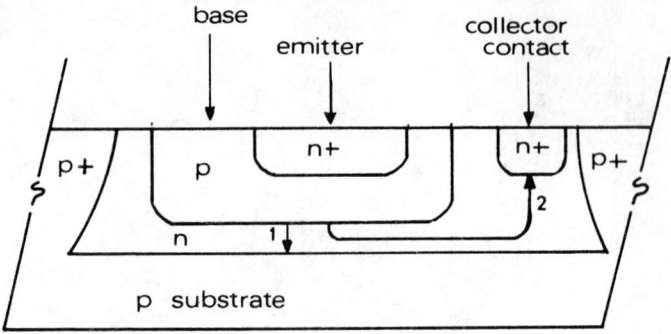

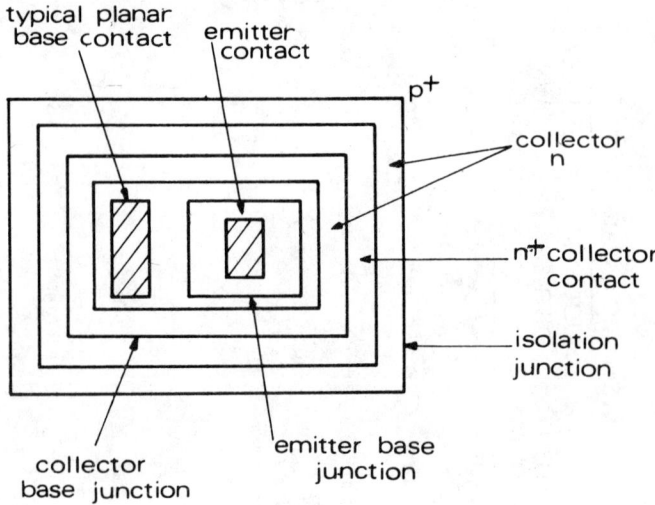

Figure 7.1  Integrated transistor.

One problem in achieving the desired result appears with surface recombination currents, and could be corrected if the emitter periphery were designed to be as minimum as possible for low-current applications. The latter statement clearly shows that geometry is an important factor in transistors. Proper surface geometry results in lower base resistance and broader operating collector current range. Figure 7.9 illustrates various types of surface geometries; Fig. 7.5 illustrates common-collector and common-collector-and-base configurations that are most commonly used in IC fabrication. Table 7.3 provides the minimum transistor surface dimensions for proper operation.

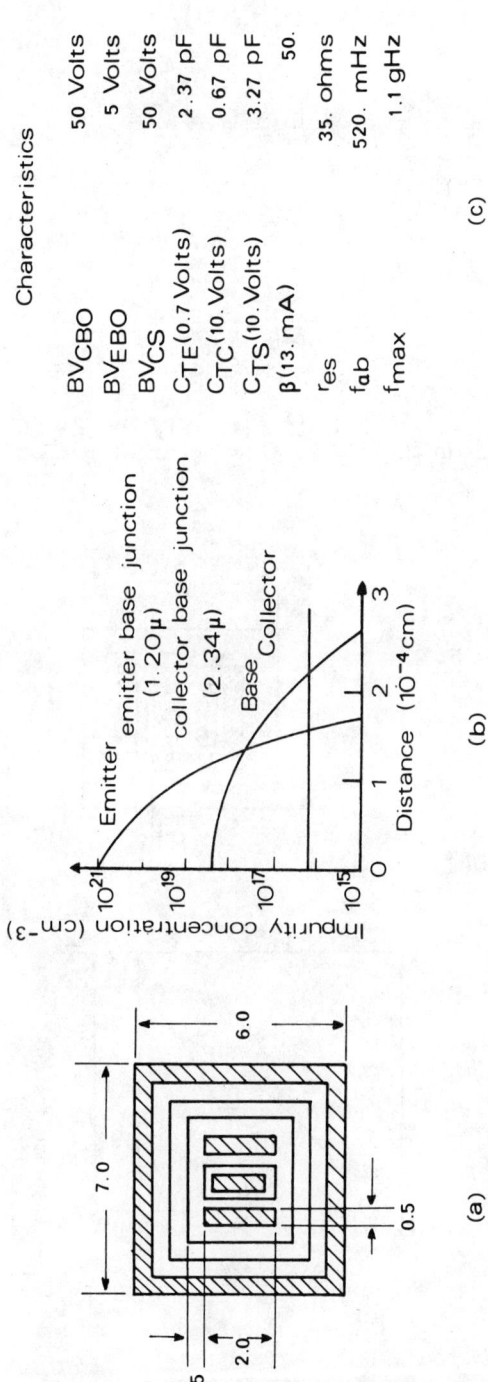

Figure 7.2  Characteristics of typical low-current Planar Transistor I.

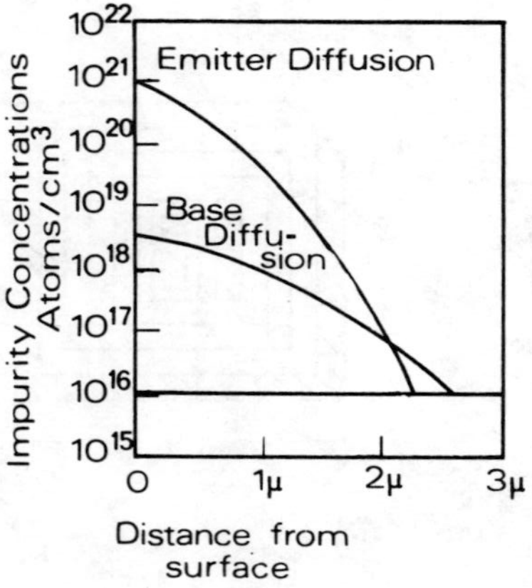

Characteristics

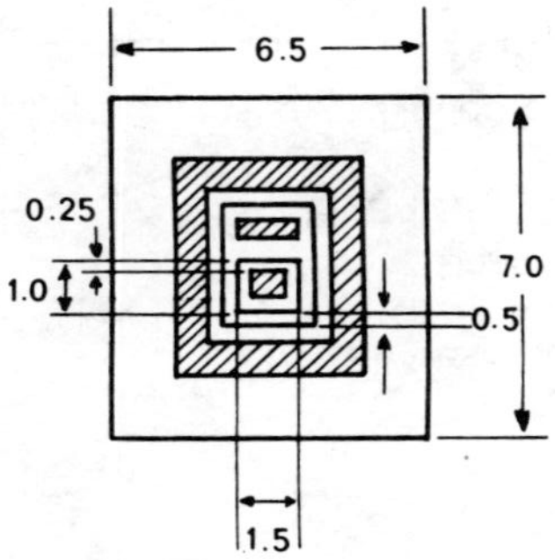

| | |
|---|---|
| $BV_{CBO}$ | 55 Volts |
| $BV_{EBO}$ | 7 Volts |
| $BV_{CEO}$ | 23 Volts |
| $BV_{CS}$ | 75 Volts |
| $C_{TE}$(forward bias) | 6 pF |
| $C_{TE}$(0.5 Volts, reverse bias) | 15 pF |
| $C_{TC}$(5. Volts) | 0.7 pF |
| $C_{TS}$(5. Volts) | 2.9 pF |
| $\beta$(10 mA) | 50 |
| $\beta$(0.1 mA) | 30 |
| $r_{CS}$ | 75 ohm |
| $V_{CE}$(SAT) | 0.5 Volts |
| $V_{BE}$(10 mA) | 0.85 Volts |
| $f_T$(5. Volts, 5 mA) | 440 mHz |
| $f_{max}$ | 1.1 gHz |

(c)

(b)

(a)

Figure 7.3  Characteristics of typical low-current Planar Transistor II.

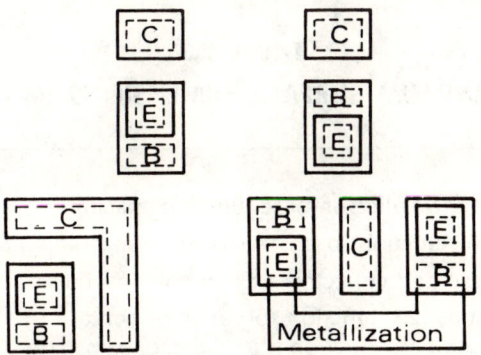

**Figure 7.4** Various transistor geometrics.

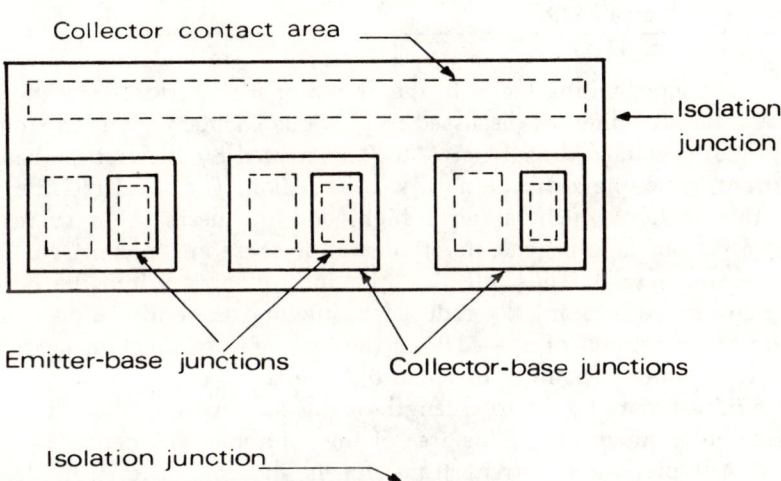

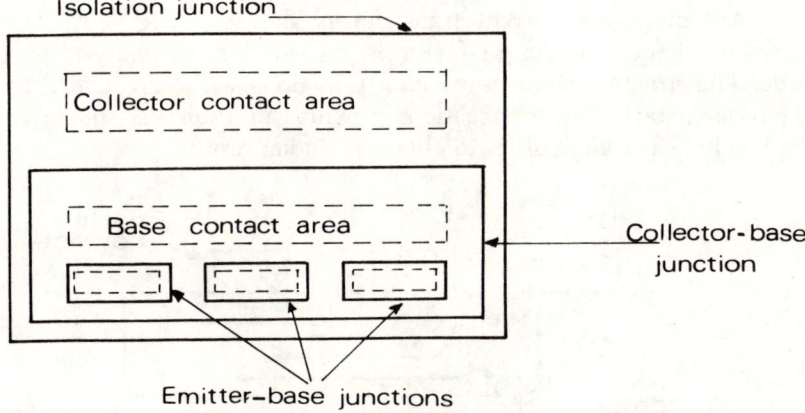

**Figure 7.5** Common-collector and common-collector-and-base transistors.

## TABLE 7.3
## MINIMUM SURFACE DIMENSIONS (MILS)

| | |
|---|---|
| Emitter | 1.0 × 1.0 |
| Emitter–base to collector–base junction spacing | 0.5 |
| Emitter–base junction to emitter contact spacing | 0.25 |
| Contact (base and collector) to junction spacing | 0.5 |
| Collector–substrate junction to collector contact or collector–base junction (the collector–substrate junction extends 1.0 mil under the oxide isolation diffusion mask) | 1.0 |
| Contact area width | 0.5 |

A major limiting factor in the design of high-current transistors is emitter crowding, as discussed in previous chapters. As seen from Fig. 7.6, a voltage drop from A to B is created by the emitter–base current, which flows in the narrow base region. The sheet resistivity of this region, which is fairly high, and the decrease in current density from A to B generate an abrupt decrease in the emitter-base bias from A to B. Thus injection of minority carriers into the base region can be substantially reduced throughout the entire transistor, with the exception of the edge of the emitter-base junction nearest the base contact (point A in Fig. 7.6). Therefore, current capacity is determined more by the total length of the emitter edge adjacent to a base contact than it is by the area of the emitter–base junction.

A typical high-current transistor is shown in Fig. 7.7. Here again the important design factor is the emitter periphery-to-area ratio. The structure contains six emitter regions, each surrounded by the base contact. The specifications provided in Table 7.3 should be applied here as well in order to obtain optimum results.

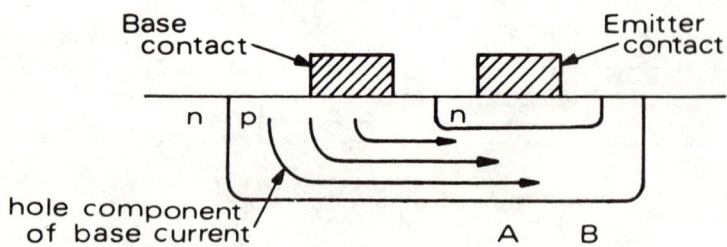

Figure 7.6  Emitter crowding.

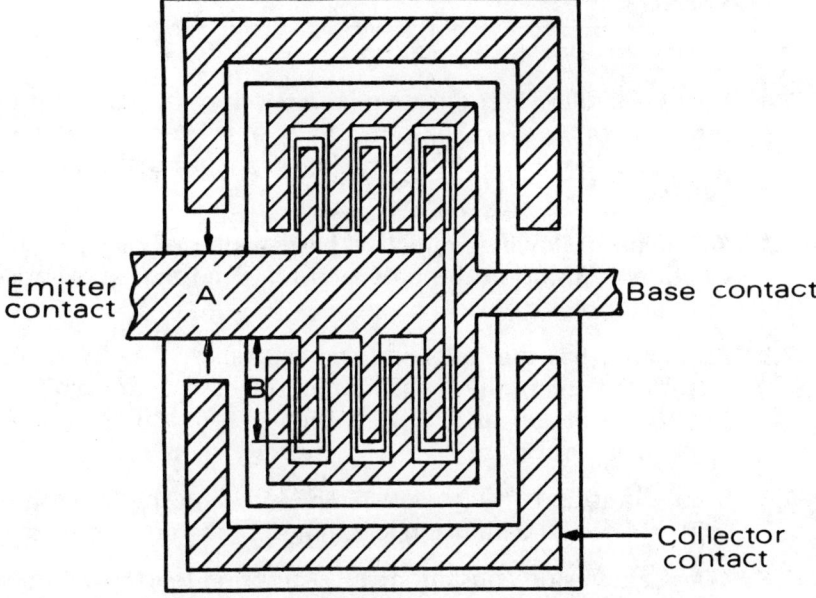

Figure 7.7　High current transistor surface geometry.

## Exercises

**7.1** Given the following: effective base width of 0.5 cm, emitter resistivity of 0.1 Ω-cm, base resistivity of 0.5 Ω-cm, and electron diffusion length in emitter of 4 μm. Find the emitter efficiency.

**7.2** Given the following: effective base width $W$ of 0.53 μm, electron diffusion in the base of 4 μm. Find the base transport factor of the device.

**7.3** An emitter has the following characteristics: area of $2(6.45 \times 10^{-6})$ cm²; impurity concentration of base under emitter, $5 \times 10^{17}$; voltages of 1, 6, 12 V. Find the transition capacitance for each voltage.

**7.4** If the diffusion constant is 0.05 cm²/sec, and the lifetime is 0.005 sec, find the diffusion length in centimeters.

**7.5** Given a diffusion constant of 17, a diffusion length of 5 mils, a junction area of 4, and $p_n$ of 2, calculate the reverse current.

**7.6** Given a reverse current of 5 mA and a forward current of 8 mA, calculate the forward voltage.

**7.7** Given the density of electron donor atoms as 5 and $\varepsilon_{max}$ as $2 \times 10^5$ V/cm, calculate the breakdown voltage.

**7.8** If the length of a resistor is 7 mils and its width 0.5 mil, find the end effect correction factor.

**7.9** Name several characteristics for optimum transistor design.

**7.10** Briefly describe emitter crowding.

# chapter eight
# Photoengraving and Mask Fabrication

## 8.1 Introduction

One of the most important fabrication steps is photoengraving and mask fabrication. Once the circuit has been designed and its theoretical operation has been tested either by breadboarding or by computer, it enters the production route, passing through artwork, digitizing, photoengraving, mask fabrication, and other steps. It is very important for the design engineer to be familiar with all these production steps.

## 8.2 Artwork

The first fabrication step, prior to photoengraving and mask layout, is the artwork. The theoretical design of the circuit is changed into a large layout, and each element on that layout accurately represents the size and location of all the components. The layout is normally in the form of a drawing on coordinate paper, which is then reproduced in a "cut-and-peel" red- or amber-colored plastic on a Mylar base, available under the brand names Rubylith or Studnite.

Freehand cutting is normally capable of obtaining an accuracy of better than $\frac{1}{20}$ in., which is usually adequate for most integrated circuits.

## 8.3 Mask Design

The fabrication of an integrated circuit is composed of several artwork drawings and, naturally, several masks. Each mask has the function of defining specific areas as follows:

1. Preepitaxial growth: placing of an $n+$ buried-layer diffusion to reduce collector series resistance.
2. Isolation diffusion: generating separate $n$ regions in the epitaxial layer for transistors, diodes, resistors, and capacitors.
3. Base diffusion: forming the base region of transistors, one side of a diode, and resistor areas.
4. Emitter diffusion: placing the transistor's emitters, capacitor regions, low-resistance crossovers, and low-resistivity terminal areas on $n$-collector regions.
5. Preohmic: making holes through the surface oxide to permit the metallization to make contact with terminal areas of circuit elements.
6. Metallization: establishing the metal interconnection pattern to wire the circuit and provide pads to which external connections can be made.

The integrated circuit masks are produced by a photographic process. They normally contain an array of images, which may or may not be similar in shape or size. The process control requirement of IC fabrication requires that control images be fabricated within the array. Large reductions in going from original artwork to final mask, on the order of 200 times, are common in microelectronics. These reductions must usually be obtained in two steps. Overall image accuracy in the camera system of 50 $\mu$in. is possible if careful control is exercised. This is far beyond the resolution capability of any of the commercially available photoresist materials. The single image photographed in the first reduction must be repeated to produce the array required. This can be accomplished by any of a number of different step-and-repeat systems. The same result may be attained by using either multilens techniques or a pinhole camera system.

The final reduction is made by using high-resolution photographic plates. Image quality is improved with a high-quality lens and a monochromatic light source. Image quality is limited by the resolving power of the lens and brightness ratio of the final detail to the large areas.

## 8.4 Photoreduction

The original artwork has to be substantially reduced in order to be transferred onto the glass mask, and this is accomplished by microphotography. The amount of reduction accomplished is proportionate to the distance of the subject from the lens to the focal length of the lens, or

$$\text{reduction ratio} = \frac{\text{lens to subject distance} - \text{focal length of lens}}{\text{focal length of lens}}$$

Thus a 4-in. lens at an 80-in. distance from the subject would give about a 20× reduction. The maximum practical ratio is about 50 : 1 for one-step reduction. Higher reductions of, say, 200× require a two-step process. Several important factors have to be taken into consideration during reduction; these are aberration, diffraction, and exposure time, and they are dependent on one another. For instance, a full lens aperture will create a maximum aberration degradation, whereas at minimum aperture aberration is practically nonexistent, but degradation by diffraction occurs. General practice is to step down a lens one or two stops from its maximum aperture.

The resolution of a lens is its ability to project a fine pattern; it is expressed in number of lines per millimeter. A test pattern for measuring resolution is made up of alternating equally wide black and clear lines. The sum of the widths of a single black and adjacent clear line is considered a line. The resolution limit is given by the equation

$$\frac{\text{resolution limit in lines}}{\text{mm}} = \frac{10^6}{\lambda \, (f\text{-number})}$$

where $\lambda$ is the wavelength in millimicrons (m$\mu$).

Table 8.1 provides information on the minimum focal length for best resolution.

## 8.5 Masks

The masks actually used in integrated-circuit processing are developed photographic emulsions deposited on glass plates. Since we previously talked briefly about mask design, it would be appropriate

**TABLE 8.1**

| CHIP SIZE (IN.) | ELEMENT SIZE | ARTWORK TIMES | ARTWORK DIMENSION (MILS) | FIRST RED. ($R$) | FOCAL LENGTH $f$ (IN.) | ARTWORK TO LENS DIST. $u = Rf$ (FT.) |
|---|---|---|---|---|---|---|
| 0.100 × 0.100 | 0.5 μm | 1000 | 100 × 100 | 50 | 14.7 | 58¾ |
| 0.050 × 0.050 | 0.5 μm | 1000 | 50 × 50 | 50 | 7.05 | 29⅓ |
| 0.050 × 0.050 | 0.5 μm | 1000 | 50 × 50 | 20 | 17.7 | 29½ |
| 0.050 × 0.050 | 0.5 μm | 1000 | 50 × 50 | 10 | 39.2 | 32⅔ |
| 0.100 × 0.100 | 0.1 mil | 500 | 50 × 50 | 50 | 7.05 | 29⅓ |
| 0.100 × 0.100 | 0.1 mil | 200 | 20 × 20 | 20 | 7.05 | 11¾ |
| 0.075 × 0.075 | 0.1 mil | 200 | 15 × 15 | 20 | 5.3 | 8⅚ |
| 0.050 × 0.050 | 0.1 mil | 500 | 25 × 25 | 50 | 3.5 | 14⅔ |
| 0.050 × 0.050 | 0.1 mil | 200 | 10 × 10 | 20 | 3.5 | 5⅚ |
| 0.075 × 0.075 | 0.1 mil | 200 | 15 × 15 | 10 | 10.6 | 8⅚ |

at this point to proceed with a practical example, which will not only give experience in mask arrangement but also in the layout of a circuit as well. Consider the amplifier of Fig. 8.1. This circuit is a linear one, but its layout and fabrication techniques apply to digital circuits as well. The first approach to the design is to lay out the circuit.

We must first designate the pin connections, which are determined by the system requirements. Figure 8.1 shows the pin designations. In the design, we see that three crossovers are required, two of which may be formed across the diffused resistors. Owing to the oxide passivation, metallization may be formed over any diffused resistor without shorting. However, the third crossing is more complicated. This type must utilize a passivated area of $n+$ diffusion for the bottom conductor, with the metalized crossover on top. This can be a special area or it can be combined with the collector region of a transistor. Thus, for design purposes of this amplifier, we make the third crossover in the collector region of Q1.

The next step determines the required isolation regions, which, for this circuit, are only two, the collector node of Q1, and the positive power supply node, which will include Q2, Q3, and all resistors. The collector of Q1 must vary with the input signal and thus must be in a separate region. The collectors of Q2 and Q3 are connected electrically to each other and to the positive supply, and thus lie in the same isolation region. Owing to the diode isolation effect, all resistors can be in the same isolation region, which is then connected

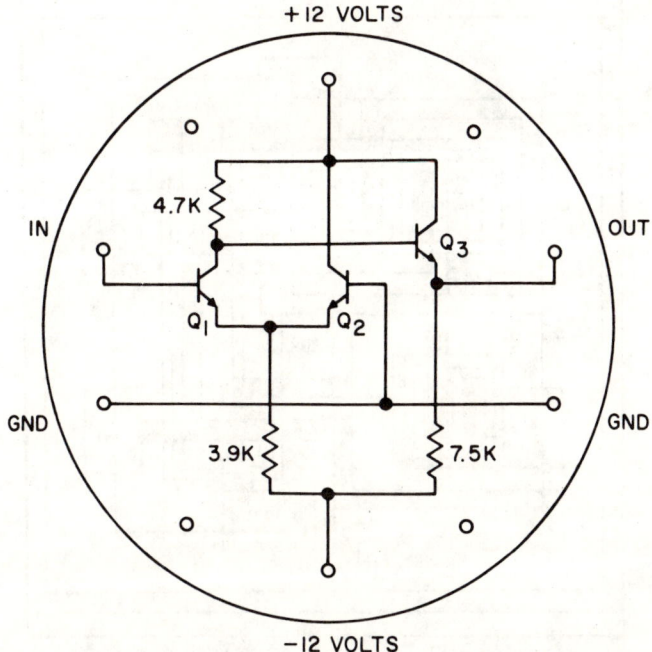

Figure 8.1  Amplifier circuit.

at the most positive potential to prevent conduction from the *p*-type resistor to the *n*-type isolation. In our example, this is the same potential as the resistor isolation region of Q2 and Q3. Thus only two isolation regions are required for the complete circuit. The resistor isolation region is connected to the +12-V supply through the collector contact of Q3. Because some of the resistors are connected to the negative supply, the *pn* junction of each resistor must withstand over 24 V of reverse bias.

To minimize the effects of parasitic capacitance and to ensure complete isolation between the various regions, the substrate should be connected to the most negative fixed potential in the circuit. In our example, this contact is obtained by metalizing to a *p*-type base diffusion under the negative supply binding pad. If negative supplies were not available, the ground potential would have been used.

Thus, on the basis of the preceding, we arrive at the layout of Fig. 8.2. A photo of a digital circuit, illustrating our requirements, is shown in Fig. 8.3.

An additional layout requirement is the type of wire bonding to be used and the size of wire required. The different wire-bonding

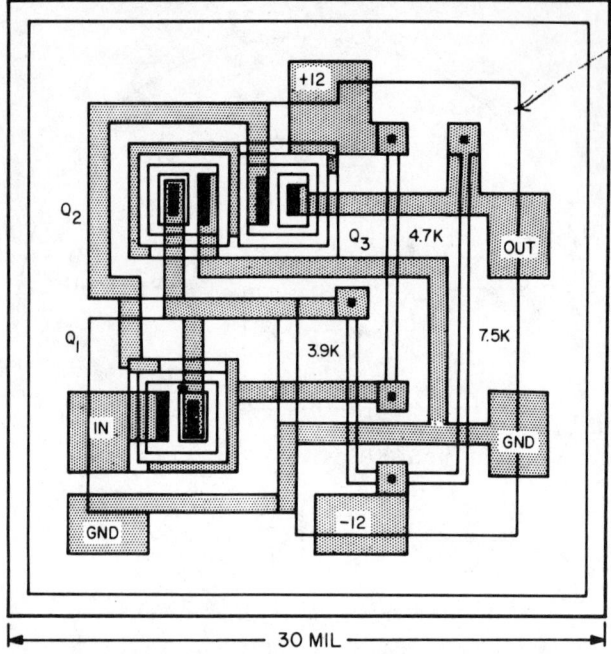

Figure 8.2  Amplifier layout.

techniques, thermal compression and ball bonding, will require different-sized bonding pads. For ball bonding, the pad should be about three times the wire diameter. For thermal compression bonding, the required area is less, and a pad approximately 2 by 3 mils is sufficient for a 1-mil wire. All pads should be placed at least 2 mils from the scribe grid to prevent possible shorting due to uneven breakage.

The following masks are required for the circuit of Fig. 8.4:

1. Channel diffusion: a deep $p$-type diffusion to isolate the active regions.

2. Base diffusion: a $p$-type diffusion to form the resistors, base of transistors, and contact area to the substrate.

3. Emitter diffusion: an $n+$ diffusion for the emitters of the transistors and the collector contacts.

4. Ohmic contact: to remove the oxide where a contact is required to the silicon.

5. Metallization: to form the aluminum interconnection pattern and bonding pads.

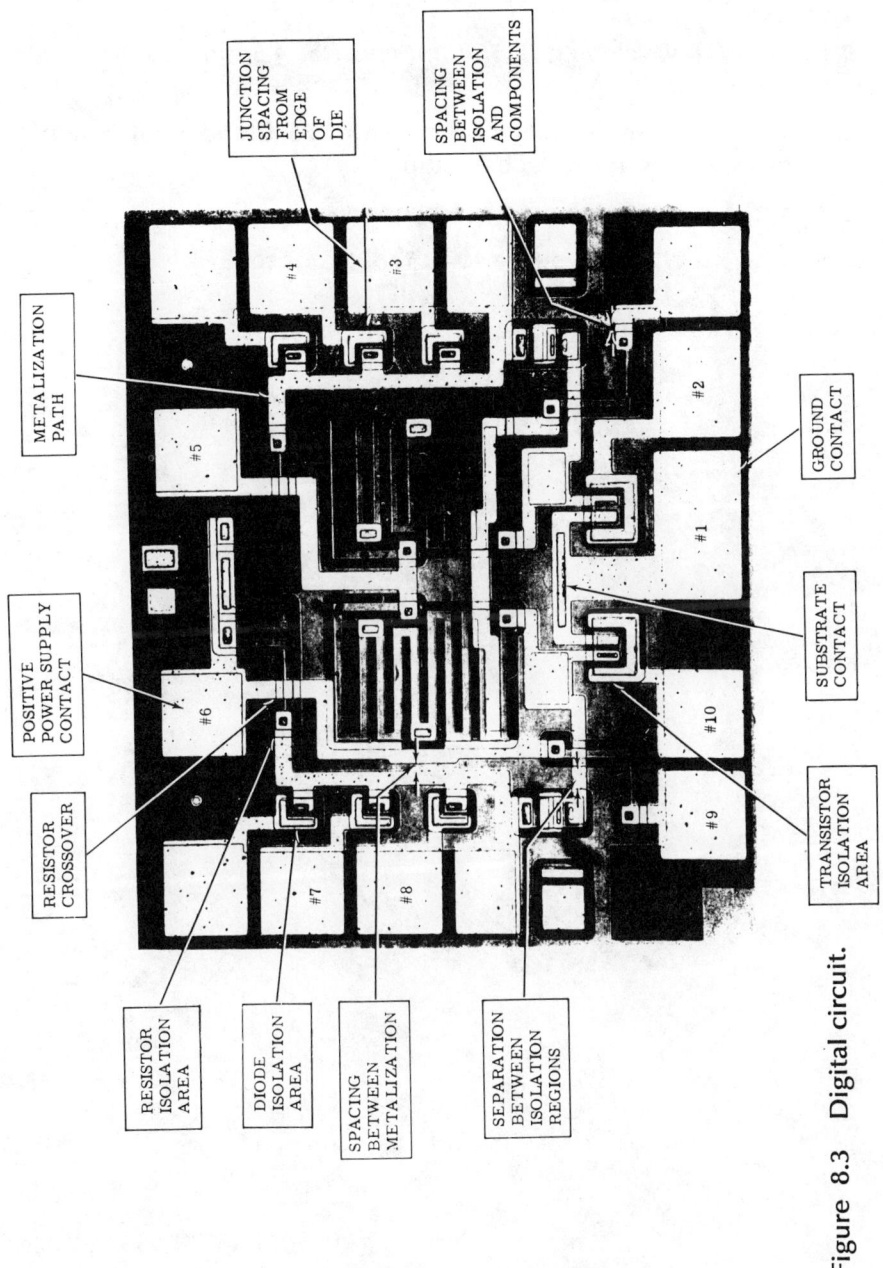

Figure 8.3 Digital circuit.

## Exercises

**8.1** Given a wavelength of 550 nm and f of 4.5, find the resolution limit.

**8.2** Given the lens-to-subject distance of 55 ft and focal length of 15 in., find the reduction ratio.

**8.3** What materials are used in artwork?

**8.4** How many steps would a 400× reduction require?

chapter nine
# Linear Integrated Circuit Design

## 9.1 Introduction

Thus far we have discussed the electrical and physical properties of the various components constituting an integrated circuit. However, the very first and most important step of the designer is the theoretical design of the actual circuit, and this varies from discrete component design. In an integrated circuit, we are not as free to employ large-value capacitors and power resistors or other components, which may be easily designed in a discrete circuit. Consideration of parasitic effects becomes more critical and is not that easy to eliminate. Therefore, the designer has to face several tolerance problems, spacing problems, and component requirements. This chapter discusses several circuit designs that have become more or less standard in IC's, as well as proper application of the various components.

Although IC's are more sensitive to the various design rules, they have solved many problems that could not have been easily faced with discrete circuits. For instance, the freedom to use a large number of active components in an IC allows one to design lower power drain circuits at much lower cost. High-frequency and high-

**104    Linear Integrated Circuit Design**

speed operation due to decreased lead length are also factors achievable with IC's.

## 9.2 Capacitor Applications

Although capacitors in an IC are costly, they do provide useful functions. Figure 9.1 illustrates a basic voltage follower circuit. Frequency compensation is not always required with this type of circuit, but resistor $R1$ and capacitor $C1$ have been included to improve stability with capacitive loading. Figure 9.2 is a complete schematic of a voltage follower with low input current. The $RC$ network, $R11$ and $C2$, is included to suppress oscillations in this feedback loop. The voltage drop across $C2$ is less than 2 V, and thus a junction capacitor can be fabricated from the emitter and base diffusions of the *npn* transistors. With this, the required capacitance can be obtained in a reasonable area of the chip without additional processing steps, as would be required if an MOS capacitor were used. The same is true for $C1$ in the same circuit.

Figure 9.3 is the integrated form of the circuit of Fig. 9.2.

## 9.3 Lateral PNP's

A lateral *pnp* transistor is shown in Fig. 9.4. The isolated $n$ region serves as the base of the transistor, with two $n+$ regions for good conductivity. The *p*-type emitter and collector regions are formed by the *npn* base diffusion, with the collector surrounding the emitter. A conceptual top view of a modified *pnp* transistor and its electronic schematic are shown in Fig. 9.5. The shaded circle repre-

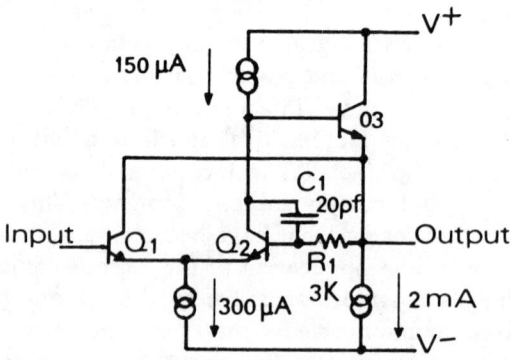

Figure 9.1   Voltage follower.

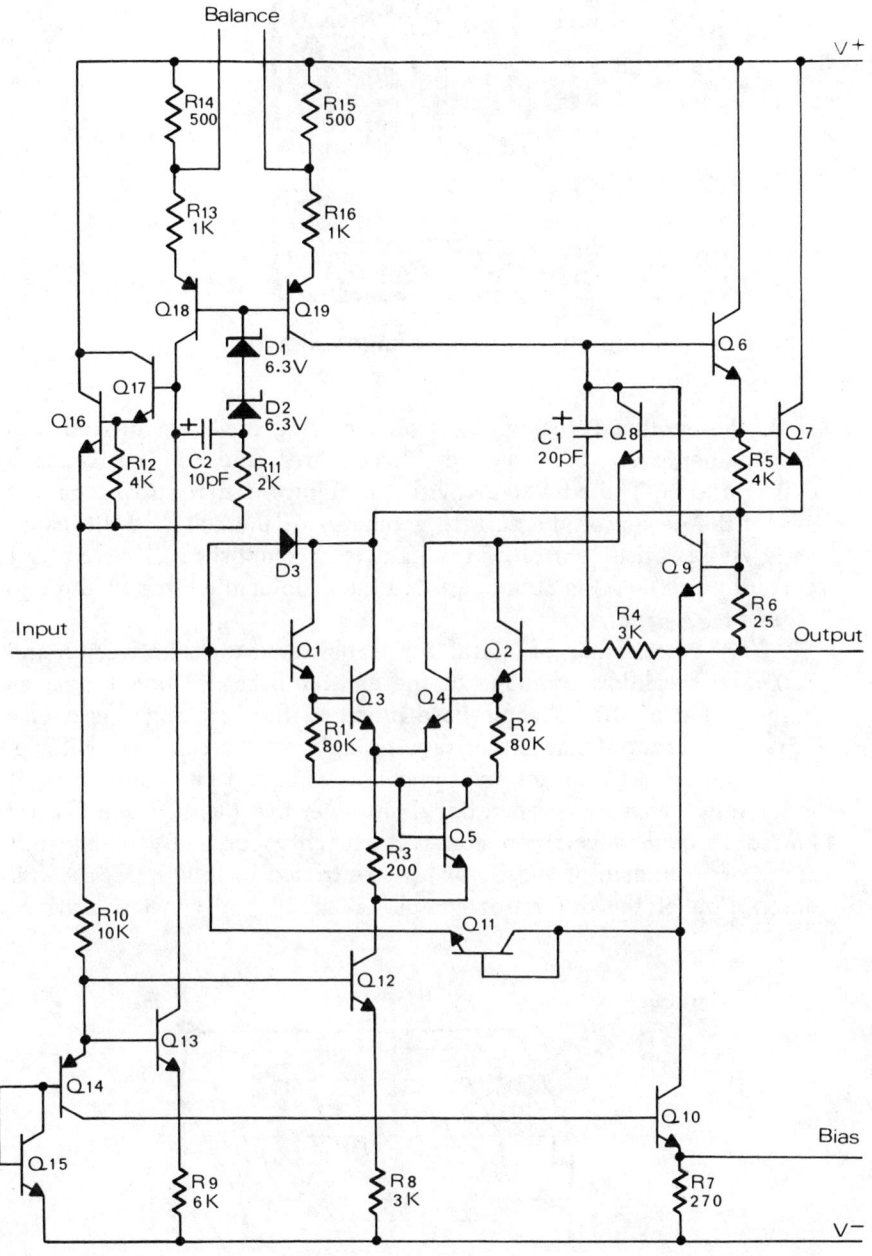

Figure 9.2 Complete schematic.

**106**  **Linear Integrated Circuit Design**

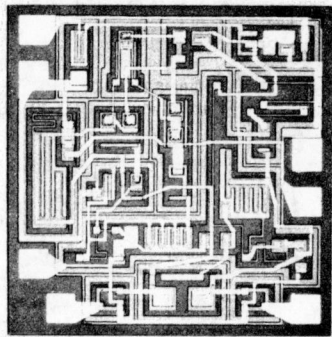

Figure 9.3  Integrated form of voltage follower.

sents the emitter diffusion. The collector ring is shown broken into two segments, $A_C$, the effective collector area, and $A_F$, the feedback collector area. The advantage of this structure is that current gain is set by the design of the surface geometry of the collector diffusion; thus very precise control of the gain is accomplished. The effective current gain ($I_C/I_B$) is about equal to the ratio of the areas of the two segments, $A_C$ and $A_F$.

The current gain of lateral *pnp* transistors ranges between 5 and 100. The breakdown voltage of the emitter–base junction is just as high as that of the collector–base junction, thus making this device attractive for input stages that require high input breakdown voltage.

Figure 9.6 illustrates an application of lateral *pnp* transistors. It is the input stage of an operational amplifier in which Q3 and Q4 are biased through a common constant current source. By using split collectors, the gain of these *pnp*'s is controlled to less than 5. A full description of the differential amplifier circuit and constant current

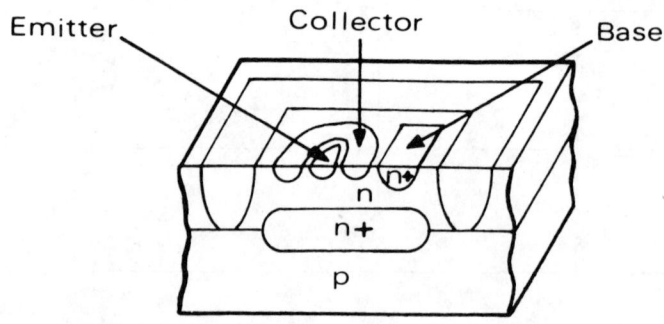

Figure 9.4  Lateral p-n-p transistor.

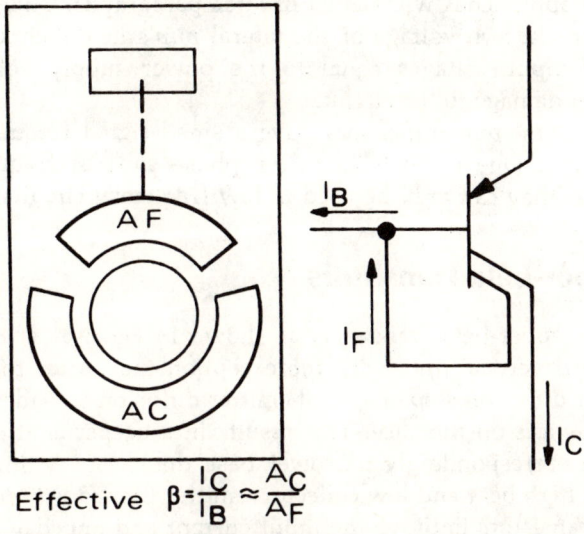

Figure 9.5  Modified lateral p-n-p transistor.

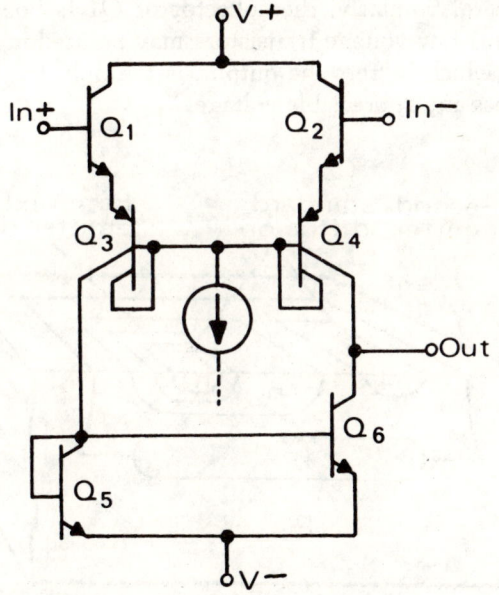

Figure 9.6  Application of lateral p-n-p transistors.

source approaches will follow in later paragraphs. The high emitter–base breakdown voltage of the lateral *pnp*'s in the circuit of Fig. 9.7 allows input voltages equal to the power supply voltage without causing damage to the circuit.

Lateral *pnp* transistors have a small-signal frequency response limitation. Due to their excessive phase shift at frequencies above 2 MHz, they can only be used in low-frequency circuits.

## 9.4 Super-beta Transistors

A super-beta transistor, as shown in Fig. 9.7, is the same as a standard vertical *npn*, with the exception that, prior to the standard emitter diffusion step, a special emitter diffusion is made for all super-beta devices on the slice. This results in a deeper emitter than usual, with a correspondingly narrower base thickness. A thin base region means high beta and low collector–emitter breakdown voltage. Super-beta transistors improve the input current and impedance of an operational amplifier.

Figure 9.8 illustrates the application of superbeta transistors. It is a typical input stage of an op-amp with very low input bias current. The diode assists in the operation of $Q2$ at a collector–base voltage near zero. Similarly, the collector of $Q1$ is bootstrapped to the output. Thus low-voltage transistors may be used in the input. Transistor $Q3$, which buffers the output, is the only transistor in the circuit that sees any appreciable voltage.

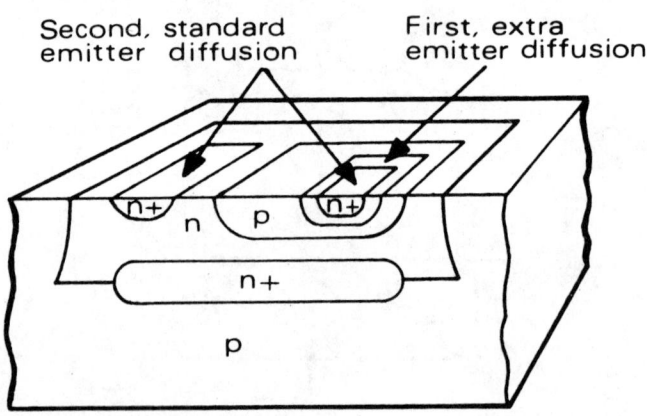

Figure 9.7  Super-beta transistor.

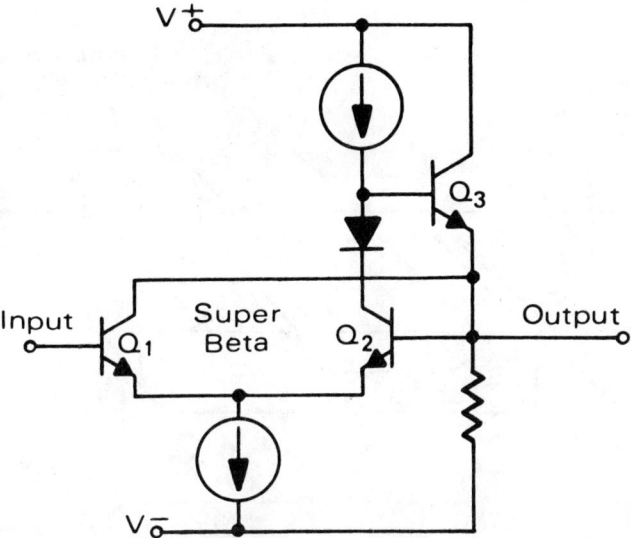

Figure 9.8 Application of super-beta transistors.

## 9.5 Schottky-Clamped Structures

A Schottky-clamped transistor and its equivalent schematic are shown in Fig. 9.9. A slightly enlarged base contact metallization is formed at the top with the darker base diffusion below, the emitter at the left, and the collector region farther down. To make the Schottky structure, a hole is simply left through the base diffusion. The junction of the base metallization with the lightly doped collector region forms the contact-barrier Schottky structure between the collector and the base, and this results in a greatly reduced turn-off time.

Schottky diodes are used to improve the speed of *npn* transistors, as shown in Fig. 9.10. A Schottky diode is simply built into the structure of the transistor, in parallel with the base–collector junction. As there are no minority carriers to store up charge in the diode structure, the switching time of the Schottky diode is extremely fast. Furthermore, the forward voltage is only about $3/10$ V. The purpose of the diode is to clamp the base to the collector, thus preventing the transistor from ever going into full saturation.

## 9.6 Pinch Resistors

A pinch resistor is a specialized type of resistor, as shown in Fig. 9.11. The illustration shows an isolated portion of the *n*-type epi layer, and, within it, a *p*-type diffused resistor. The rear portion de-

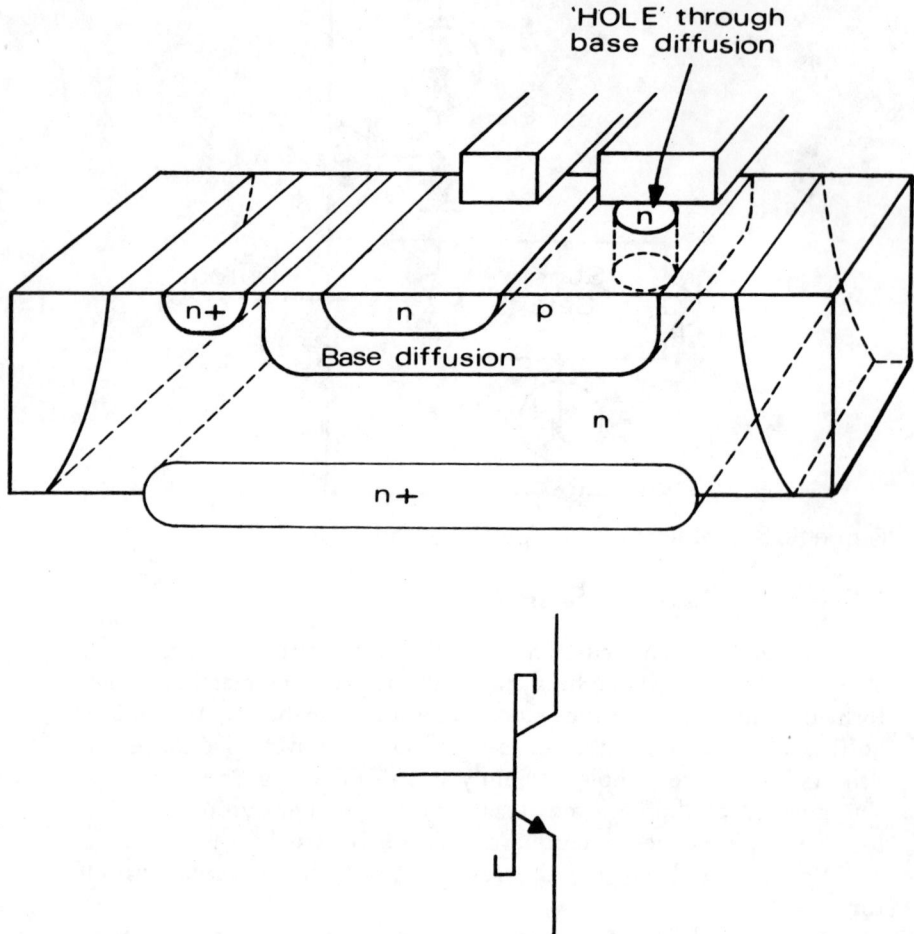

Figure 9.9  Schottky-clamped structure.

picts one of the contacts of the resistor. There is a section of $n+$ emitter diffusion over the mid-section of the resistor, which is similar to an FET gate, and the $p$-type resistor diffusion acts as an FET channel. In actual operation, a reverse bias is maintained on the *gate channel* junction, causing a constriction of the effective cross section of the diffused resistor in a sort of *pinching* action. The FET structure is not operated in the *pinch-off* region and is not used as an FET in the sense of an active device, as no signal is applied to the gate. This simply increases the sheet resistance of the resistor (10–30 k$\Omega$/

## Pinch Resistors     111

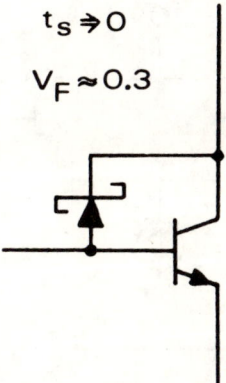

Figure 9.10   Schottky method of improving speed.

square), giving a high-value resistor in a relatively small area of the wafer.

Pinch resistors have a rather low breakdown voltage (6–8 V) due to the heavily doped gate, as well as a strong positive temperature coefficient (3000–5000 ppm/°C). An important use of pinch resistors is as bleed resistors in circuits where breakdown and tolerance problems do not affect performance. Observing again the input stage of an op-amp, as in Fig. 9.12, we see that the input has two transistors and a pinch resistor. The pinch resistor in this case acts in an

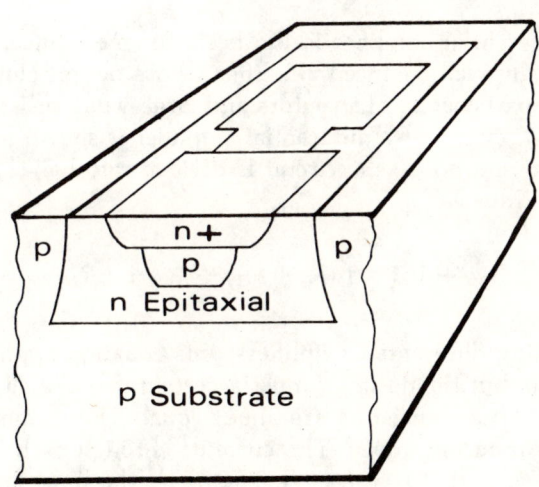

Figure 9.11   Pinch resistor.

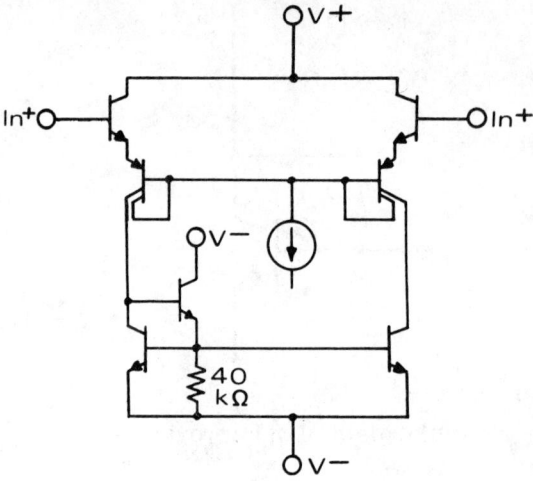

Figure 9.12   Application of pinch resistor.

emitter–base *bleeding* capacity. The input offset voltage and large-signal frequency response in this circuit are improved by the use of this resistor.

## 9.7 Differential Amplifier

The differential amplifier is the basis for most linear IC's. It is an exceptionally well balanced amplifier whose design eliminates the use of costly parts such as capacitors and large-value resistors.

The basic circuit of a differential amplifier is shown in Fig. 9.13. For dc analysis purposes, the circuit is divided into halves, as shown in Fig. 9.13b. Thus

$$V_{ee} + V_{cc} = I_e(R_e + r_{e1}) + V_{ce} + I_c R_{L1}$$

When designing differential amplifiers, this equation applies to both halves as well, but the designer must ascertain that the current gains ($h_{fe}$) of both input transistors are about equal. The common emitter resistor $R_e$ is used only once. The currents through each branch are purely additive, so that $I_1 + I_2 = I_3$. The figure shows ac input signals $e_1$ and $e_2$ applied to each side of the amplifier. This will result in amplification of the difference between the two signals, and the amplifier will reject any portion of the incoming waveform that is com-

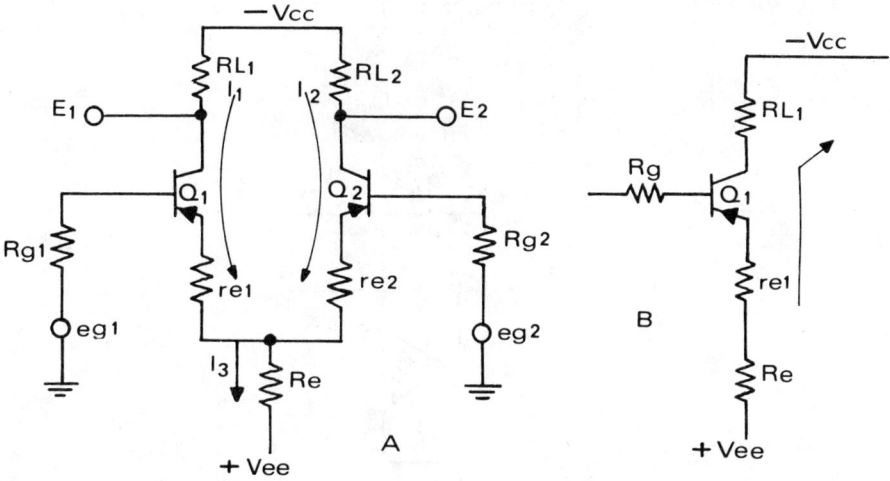

Figure 9.13 Differential amplifier.

mon to both sides. This is known as the common-mode rejection ratio (CMRR) and equals

$$E_0 \text{ (dB)} = \frac{e_1 + e_2}{A_v}$$

If the current gains are different, CMRR can be calculated as follows:

$$\text{CMRR} = \frac{2R_e(h_{fe1} \times h_{fe2})}{R_g(h_{fe1} - h_{fe2})}$$

The sections that follow provide a description of the more improved input stages of linear circuits.

## 9.8 Operational Amplifier Input Stage

Figure 9.14 shows the input stage of a linear op-amp in which transistors Q1 and Q2 are, in effect, emitter followers differentially driving the emitters of Q3 and Q4. Since Q3 and Q4 are lateral *pnp's*, they are used as common-base amplifiers, and this permits their poor frequency characteristics to be circumvented. An improved version of the basic differential amplifier is clearly shown by the addition of transistors Q5 and Q6, which form a current source and provide the load for Q4. The advantage of this configuration is that it allows the

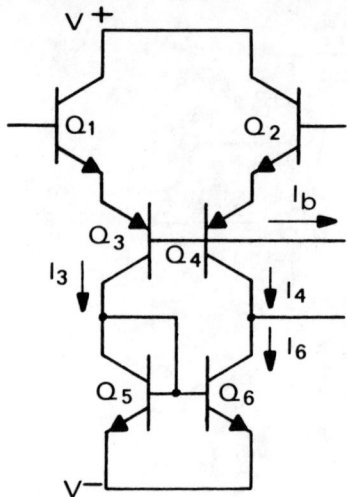

Figure 9.14   Op-amp input stage.

full differential current gain to appear single ended at the collector of Q4. Should $I4$ increase, $I3$ will show a corresponding decrease, and so will $I6$ ($I4 = I6$ in the balanced condition). The net result will be an output current equal to $I4 - I6$, and, since $I6$ always equals $I3$, the output current will be $I4 - I3$.

Additional advantages of the above circuit are as follows:

1. It provides an extremely high output impedance (about 2 M$\Omega$) to drive the second stage.
2. Since the collector of $Q4$ sees only the input impedance of the second stage as a load, the circuit provides a high-voltage gain.
3. The ac input impedance is doubled because of the two additional base–emitter junctions between the input pair.
4. Because of the uniformly doped epi base of the lateral *pnp* transistor, $BV_{CBO}$ equals $BV_{EBO}$, at a value of 80–90 V. With this type of input configuration, we can use a differential input voltage range equal to the supply voltage, without destroying the device. If conventional *pnp* transistors were used in place of the lateral ones, the input voltage would be limited to $\pm 14$ V.

## 9.9 Additional Stages

Let us further move into the second stage to which the above circuit could be connected. This stage, shown in Fig. 9.15, is a stan-

## Additional Stages 115

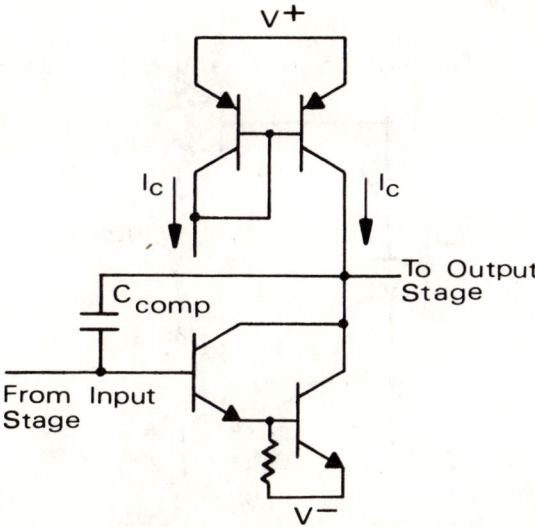

Figure 9.15 Driver stage.

dard Darlington *npn* common-emitter amplifier, with a *pnp* current source as the collector load impedance. The compensation capacitor is usually applied around this stage to obtain a constant 6 dB/octave roll off out to unity gain. The advantages of placing the feedback here are that a very high impedance on the order of 1 MΩ is seen at the input, and high current gain is available from this stage (about $2 \times 10^4$). In this manner, a fairly small capacitor (30 pF) can limit the open-loop bandwidth of the amplifier to 7 MHz.

The output stage of a complete circuit is shown in Fig. 9.16. It is a complementary emitter follower, which operates at or near class B. A small amount of standby current (100 µA) is maintained through the output pair to eliminate crossover distortion. The main advantage of this output configuration is the large amount of output current available in the positive and negative direction. However, to obtain this current gain, another *pnp* transistor is required. The following are two methods of obtaining a *pnpn* transistor at the output:

1. Use a vertical or substrate *pnp* with a base diffusion as the emitter, the *n* epi layer as the base, and the *p* substrate as the collector. This method can only be applied when the collector is directly connected to the most negative voltage in the circuit, and provides rather poor transistor characteristics.

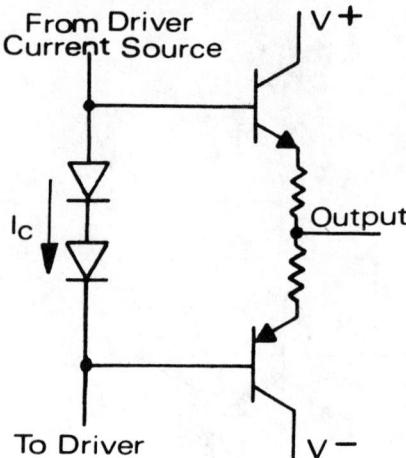

Figure 9.16  Output stage.

2. Connect a lateral *pnp* and a standard *npn*, as shown in Fig. 9.17, thus providing an equivalent *pnp* current gain equal to the product of the *pnp* and *npn* gains. The beat cutoff frequency of this device will be the same as if only the *pnp* were used.

The standby current for an output stage is usually developed by inserting either a resistor or diffused junctions (diodes, base–emitter junctions) between the bases of the output-pair transistors and in series with the collector load of the driver. The use of resistors is not normally recommended because of poor thermal matching as

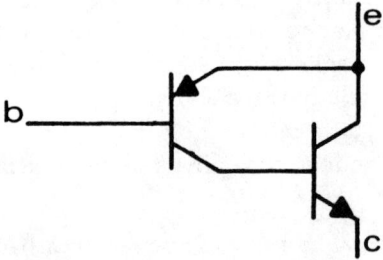

Figure 9.17  Use of a lateral p-n-p transistor in conjunction with a vertical n-p-n device provides an equivalent p-n-p current gain which is equal to the product of the p-n-p and n-p-n current gains.

compared to diodes. Active diodes ($V_{CB} = 0$) or transistors are generally used to develop the necessary turn-off voltage required for the proper standby current.

## 9.10 Biasing Circuits

Modern op-amps are not restricted to a narrow range of supply voltages. Most amplifiers can operate over a 5 : 1 supply voltage range, and this provides the designer with great flexibility, thanks to current sources applied as collector loads. For a given input current, the output current will be maintained for output voltages from near saturation to $BV_{CEO}$; the output impedance will equal $h_{oe}$ (output conductance in common-emitter configuration).

Various biasing methods are described here, and the first is shown in Fig. 9.18. A constant collector current is maintained through Q19 while the supply voltage is varied over a wide range. A fairly constant current through Q19 is obtained because the $V_{be}$ of Q18 changes by only 18 mV/octave of the collector current. For a 4 : 1 change in supply voltage, the collector current of Q18 will also change by 4 : 1 (since almost all the current through R1 becomes the collector current for Q18). R1 can be fabricated in the form of a buried FET resistor, and thus its value will be voltage dependent, increasing with increasing voltage. Therefore, the actual change in Q18 collector current will be less than the supply voltage change.

Transistor Q19 biases Q18 and also supplies current to R9. Assuming that the operating current is 100 μA, the $V_{be}$ of Q18 will

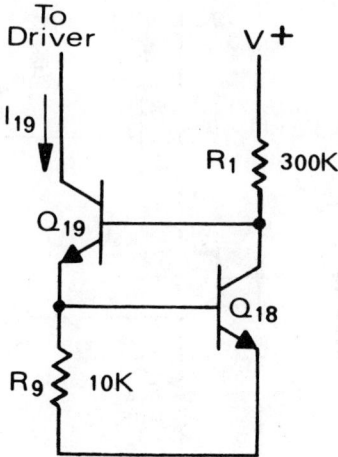

Figure 9.18  Biasing methods.

be about 600 mV, and current through $R9$ will be about 60 μA. Thus, when the $V_{be}$ of $Q18$ changes by 36 mV for a 4 : 1 change in supply voltage, the current that $Q19$ has to supply to $R9$ changes by only 6 percent. $Q19$ also supplies base current to $Q18$, but this is small compared to the current supplied to $R9$. The large value of $R1$ would appear to require a large die area, but, in fact, $R1$ can be fabricated as a buried epi FET functioning as a resistor and consuming less die area. The collector FET resistor was purposely not mentioned earlier in the chapter, so that it could be described in a paragraph closer to its application.

A collector FET resistor, as shown in Fig. 9.19, is used to achieve high-value resistances with the least die area consumed. It is in effect a simplified pinch resistor, using an isolated strip of $n$-epi material as the channel, rather than a separately diffused $p$-type channel. The mid-section of the epi strip is covered over by a $p$-type $npn$ base diffusion. The term *collector FET* derives from the fact that the epi region is usually used as the collector for $npn$ transistors. The advantage of this device is its high sheet resistance per square and its high breakdown voltage.

The open-loop biasing approach is shown in Fig. 9.19. Here $Q20$ has a constant emitter current supplied to it by $Q19$. Due to the common process and existing temperature conditions, the current gain of $Q20$ will track the current gains of $Q3$ and $Q4$ of Fig. 9.15. This is accomplished by using $Q21$ and $Q22$ as a current source

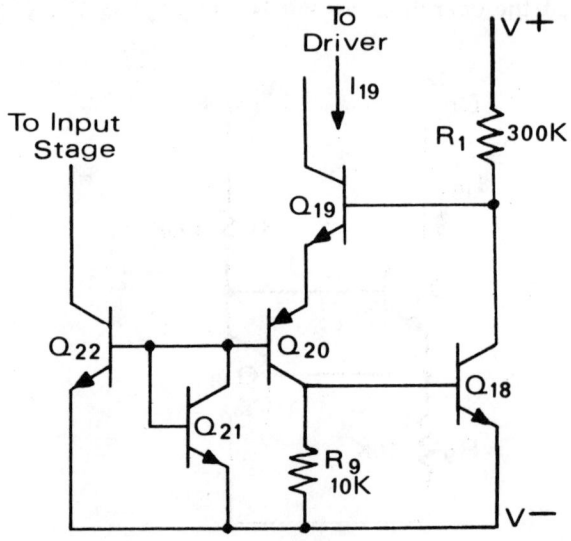

Figure 9.19 Collector FET resistor.

to transfer the base current of Q20 to the base of Q3 and Q4 (Fig. 9.15). This configuration requires that Q20, Q3, and Q4 operate in a tracking function.

Finally, a closed-loop approach is shown in Fig. 9.20. The circuit is designed so that $I_{10}$ equals the sum of the input pair (Q1, Q2) collector currents. If and when, for example, temperature increases, less base drive ($I_{3-4}$) is required to maintain $I_3$ and $I_4$ constant. Therefore, if $I_9$ does not change, $I_3$ and $I_4$ will increase. However, $I_3 + I_4 + I_{3-4}$ passes through the reference diode Q8, and, as the current through this diode increases, it causes an equivalent increase in the collector current of Q9. Increasing $I_9$ will then reduce the base drive to Q3 and Q4, restoring the original currents. The equations that follow verify these statements:

$$I_{10} = I_{3-4} + I_9 \quad \text{and} \quad I_9 = I_3 + I_4 + I_{3-4}$$

since the collector current of Q9 equals the collector current of Q8.

then $\qquad$ If $I_3 = I_4 = I_c$ and $I_{3-4} = (2I_c/h_{fe})$

$$I_{10} = 2I_c \left(1 + \frac{2}{h_{fe}}\right)$$

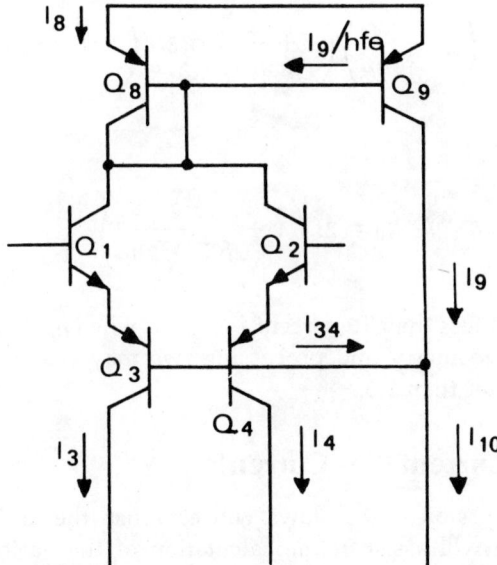

Figure 9.20 Closed-loop approach.

The preceding discussion does not mention the current gain of $Q9$. In reality, the base current of $Q9$ also contributes to the collector current of the first stage (which means that not all this current flows through $Q8$). This modifies the solution slightly. Assuming that the $h_{fe}$ of $Q9$ is about $h_{fe} = Q3 - Q4$, because of the identical processing and geometries, and that the $V_{be}$ matching of $Q8$ and $Q9$ always maintains equal emitter currents, then

$$I_8 + \frac{I_9}{h_{fe}} = I_3 + I_4 + I_{3-4}, \qquad I_9 + \frac{I_9}{h_{fe}} = I_8$$

and $I_{10} = I_9 + I_{3-4}$ as before.

We can simplify the above as follows:

$$I_3 = I_4 = I_c \quad \text{and} \quad I_{3-4} = \frac{2I_c}{h_{fe}}$$

then

$$I_9\left(1 + \frac{2}{h_{fe}}\right) = 2I_c + \frac{2I_c}{h_{fe}}$$

$$I_9 = I_{10} - I_{3-4} = I_{10} - \frac{2I_c}{h_{fe}}$$

$$\left(I_{10} - \frac{2I_c}{h_{fe}}\right)\left(1 + \frac{2}{h_{fe}}\right) = 2I_c\left(1 + \frac{1}{h_{fe}}\right)$$

Solving for $I_{10}$,

$$I_{10} = 2I_c\left(1 + \frac{2}{h_{fe}^2 + 2h_{fe}}\right)$$

For an error of less than 10 percent $I_{10} = 2I_c$, when $h_{fe} > 3.5$.
With proper geometry and processing, we may easily obtain a current gain greater than 3.5.

## 9.11 Direct-Current Bias Currents

The discussion that follows will establish the dc bias currents, which in turn will assist in the calculation of the various gains and bandwidth. Let us examine the complete schematic of an operational amplifier, as shown in Fig. 9.21. The bias current of the amplifier is determined by $R_4$ and $R_5$. At a supply voltage of $\pm 15$ V, the current through $R_5$ is given by

$$I_5 = \frac{V^+ + V^- - V_{be12} - V_{be11}}{R_5}$$

where $V^+ = V^- = 15$ V, $V_{be11} = V_{be12} = 0.6$ V, and $R_5 = 39$ k$\Omega$. Since $Q12$ is the reference diode for the driver current source $Q13$, and assuming that the $h_{fe}$ of $Q13$ is about 5, the collector current of $Q13$ will be about $I_5 - (I_5/h_{fe})$ or 600 $\mu$A.

Current $I_{10}$ must be calculated graphically or by successive approximations, and this will be one of the exercises at the end of the chapter. The designer of the circuit does not have this problem since $I_{10}$ is known to him, and his only problem is finding the value of $R_4$ (about 2800 $\Omega$).

The graphical approach of the solution for $I_{10}$ is given here, but it is left up to the student to actually plot it. On a sheet of three-cycle semilog graph paper, plot $\Delta V_{be}$ from 0 to 180 mV along the linear axis and $I_c$ from 1 to 1000 $\mu$A. Draw a straight line from the coordinate 0 mV–1 $\mu$A to 180 mV–1000 $\mu$A (60 mV/decade). This line represents the movement of $V_{be} - V_s - I_c$ for $Q11$. Now plot a curve of $(\Delta V_{be} + R_4 I_c) - V_s - I_c$ for $Q10$. For example, at $I_c = 1$ $\mu$A,

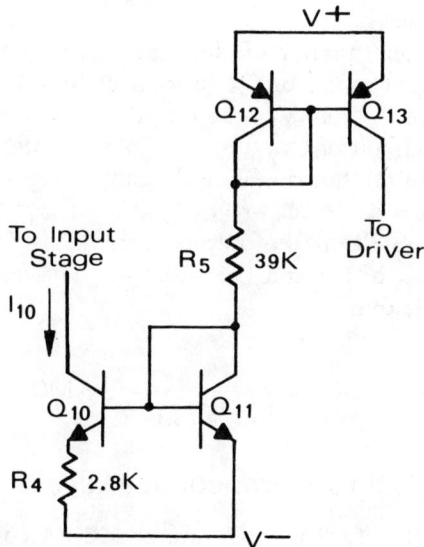

Figure 9.21 Complete schematic of op-amp.

$$V_{be} = \Delta V_{be} + R_4 I_c = 0 \text{ mV} + 2.8 \text{ k}\Omega \, (1 \, \mu\text{A}) = 2.8 \text{ mV}$$

At $I_c = 10 \, \mu\text{A}$,

$$V_{be} = \Delta V_{be} + R_4 I_c = 60 \text{ mV} + 2.8 \text{ k}\Omega \, (10 \, \mu\text{A}) = 88 \text{ mV}$$

This procedure is repeated for 2, 5, 20, 50, and 100 $\mu$A. Using a french curve, connect the points calculated. Draw a line along the $I_c = 740 \, \mu$A line until it meets the Q11 line. Draw another line perpendicular to the 740-$\mu$A line, joining it at the intersection of the Q11 line (about 172 mV). Continue the 172-mV line until it reaches the Q10 + $R_5$ line. Read the current ($I_c$) at the intersection (about 27 $\mu$A).

We now have established that the driver is operating at 600 $\mu$A, and the input stage is operating at about 27 $\mu$A (13.5 each side).

## Input Stage Gain

The gain ($g_m$) of the input stage is determined by its operating current and $s$ is given by

$$g_m = \frac{I_c}{V_{be}} = \frac{qI_c}{kT} \left( \frac{kT}{q} = 26 \text{ mV} \right)$$

However, only one quarter of the input signal voltage ($V_i$) appears across either input of Q3 or Q4 (a total of four $V_{be}$ junctions appear across $V_i$). Since $V_i = 4 \, \Delta V_{be}$, the output signal current for each side is ¼ $g_m V_i$. The circuit of Q5, Q6, and Q7 translates Q3 signal current to the output of Q4, thus doubling the output signal current. The output current ($I_o$) is $\Delta I_3 + \Delta I_4$ or $(1/1) g_m V_i$. At $I_c = 13.5 \, \mu$A, $g_m = 500$ microsiemens ($\mu$S); thus the output current $I_o = V_i$ (250 $\mu$S). At 13 $\mu$A for $I_c$, the $h_{oe}$ of Q4 and Q6 is 0.2 $\mu$S. The output impedance of the input stage is thus

$$R_{o1} = \frac{1}{h_{oe4} + h_{oe7}} = 2.5 \text{ M}\Omega$$

## Driver Stage Gain and Impedance Levels

Knowing that Q17 is operating at 600 $\mu$A collector current (almost all the collector current of Q13 flows through Q17), and that Q16 is operating at 15 $\mu$A [($V_{be17}/R_{12}$) + $I_{b17}$], and assuming that the current gains for Q17 and Q16 are 150 and 100, respectively, the input impedance may be calculated as follows:

$$R_{in2} = (h_{fe16} \times h_{fe17})(R_{11} + r_{e17})$$

If

$$r_{e17} = \frac{26 \text{ mV}}{0.6 \text{ mA}} = 43 \ \Omega \text{ and } R_{11} = 50 \ \Omega$$

then

$$R_{in2} = 1.40 \text{ M}\Omega$$
$$R_{o2} = \frac{1}{h_{oe17} + h_{oe13}}, \quad R_{L2} = h_{fe14} R_L$$

and

$$R_{o2} = 125 \text{ k}\Omega, \quad R_{L2} = 300 \text{ k}\Omega$$

where

$$h_{oe17} = 4 \ \mu\text{S}$$
$$h_{oe13} = 4 \ \mu\text{S}$$
$$R_L = 2 \text{ k}\Omega$$
$$h_{fe14} = 150$$

For a total collector impedance ($R'_{L2}$) of $R_{o2}//R_{L2} = 88.5$ k$\Omega$,

$$A_{v2} = \frac{R'_{L2}}{r_{e17} + R_{11}} = 950 = 59.6 \text{ dB}$$

Having calculated the input impedance ($R_{in2}$) of the driver stage, we can determine the voltage gain ($A_{v1}$) of the input stage as follows:

$$V_{o1} = I_o(R_{o1}//R_{in2}) \quad \text{and} \quad I_o = \tfrac{1}{2} g_m V_i$$
$$A_{v1} = \frac{V_{o1}}{V_i} = \frac{1}{2} g_m (R_{o1}//R_{in2}) = 225 = 46.1 \text{ dB}$$

The values of $g_m$, $R_{o1}$, and $R_{in2}$ are already known from previous calculations. The overall amplifier gain is

$$A_v = A_{v1} + A_{v2} \quad \text{or} \quad 59.6 + 46.1 = 105.7 \text{ dB } (172{,}000)$$

## Calculation of Open-Loop Bandwidth

Refer to Fig. 9.22 in which the amplifier is treated as a building

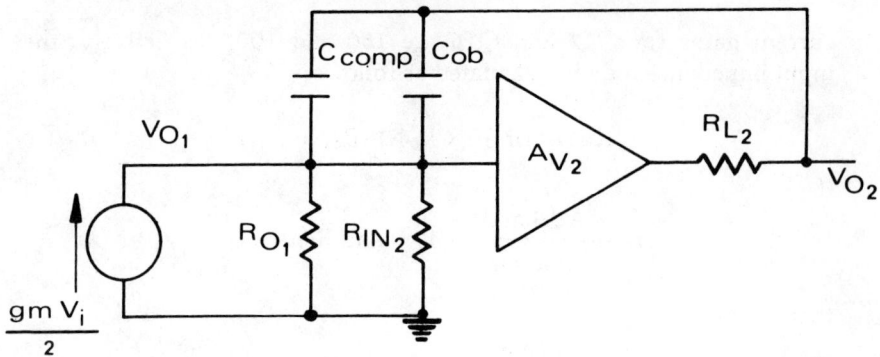

Figure 9.22 Building block of amplifier.

block. The bandwidth is determined by the gain and input impedance of the driver, the output impedance of the input stage, and the value of the compensating capacitor used. Thus the bandwidth may be calculated as follows:

$$BW(f) = \frac{1}{2(A_{v2} + 1)(C_{ob} + C_{comp})(R_{o1} + R_{in2})}$$

where

$$A_{v2} = 950$$
$$C_{ob} = 2.5 \text{ pF}$$
$$C_{comp} = 30 \text{ pF}$$
$$R_{o1} = 2.5 \text{ M}\Omega$$
$$R_{in2} = 1.4 \text{ M}\Omega$$

Therefore, $BW(f) = 5.8$ Hz. Knowing BW and $A_v$, the 0-dB gain frequency may be obtained as follows:

$$f(0 \text{ dB}) = BW(A_v) \quad \text{or} \quad 5.8 \text{ Hz}(1.72 \times 10^5) = 1 \text{ MHz}$$

The preceding discussion has provided a complete description of a linear integrated circuit analysis and can be used as a guide for other linear circuit calculations.

## Exercises

**9.1** Describe some of the functions of a capacitor in an IC.

**9.2** Briefly describe the collector FET resistor and its function.

**9.3** A particular amplifier design requires improved input current and input impedance. What sort of transistor would you use?

**9.4** To receive an increased current gain, what sort of circuit should you employ?

**9.5** Refer to Section 9.11, and actually plot $I_{10}$ on semilog paper (any office supply or engineering supply store should carry this paper).

**9.6** Starting from Section 9.7, collect the individual circuits discussed and put them together into a complete circuit.

**9.7** What are the advantages of Schottky diodes?

**9.8** Refer to the circuit shown. Given the following: $+V_{CC} = -V_{CC} = 12$ V, $h_{fe}$ of each transistor $= 50$, $V_{be}$ of each transistor $= 0.7$, and $h_{oe}$ of each transistor $= 0.4$ $\mu$S. Find the total voltage gain, the open-loop bandwidth, and the input stage gain.

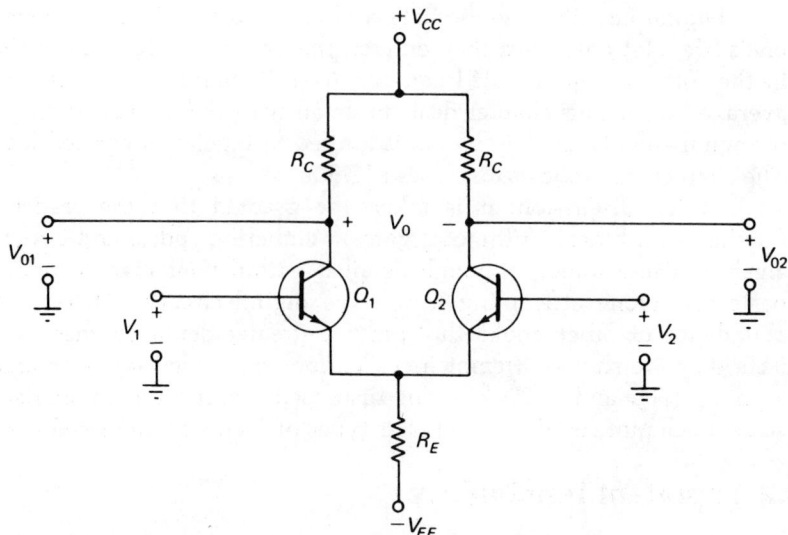

## chapter ten
# Digital Circuit Design

## 10.1 Introduction

Digital circuits have become a very important factor in everyone's life. Not only have they entered the various professional fields, in the form of sophisticated equipment and computers, but also the average home. This chapter deals in detail with the important subject of digital circuits, but discussion is limited to bipolar integrated logic. The chapter that follows discusses MOS logic.

In the discussion, it is taken for granted that the reader is familiar with binary arithmetic, gates, numbering codes, and Boolean algebra. These topics, although detailed within their own scope, are basic to anyone attempting to analyze digital circuits. There is an abundance of other books that provide greater detail on these subjects than we should attempt here. Various outmoded types of logic, such as RTL and DTL, are not discussed, except for comparison, since much more modern and faster types of logic have replaced them.

## 10.2 Important Terminology

Prior to the discussion of the various digital integrated families, the reader should become familiar with various important terms and

## Important Terminology

parameters that are constantly used and greatly influence the design of digital circuits.

## Positive and Negative Logic

The various terms used in the industry to describe the inputs and outputs of logic circuits are 1 (or high or true) and 0 (or low or false). Furthermore, there are two types of logic systems available, positive and negative. For positive logic, 1 represents a "high" voltage, and 0 represents a "low" voltage. For negative logic, 1 represents a "low" voltage, and 0 represents a "high" voltage. It has become customary to consider "low" voltages as the most negative (or least positive) and "high" voltages as the most positive (or least negative).

Although the various logic families, such as TTL and ECL, have not been discussed as yet, let us use these two families to understand more clearly the meaning of positive and negative logic. A TTL circuit has been traditionally designed with the NAND gate as the basic logic function. However, the basic function in ECL is the NOR gate (positive logic). Thus the designer may either design ECL systems with positive logic, using the NOR, or with negative logic, using the NAND. The question now is which is the most convenient system. In reality, conversion from one system to the other is not such a complicated affair. It is actually left up to the designer to use the logic form with which he is familiar, bearing in mind of course the definition of levels as described next.

Binary logic must have two states to represent the binary 1 and 0. For example, with ECL, the typical states are a high level of $-0.9$ V and a low level of $-1.7$ V. Two choices are then possible to represent the binary 1 and 0. Positive logic defines the 1 or high state as the most positive voltage level, whereas negative logic defines the most negative level as the 1 or high. Due to the difference in the definition of states, the basic ECL gate is a NOR function in positive logic and a NAND function in negative logic.

Figure 10.1 more clearly shows this comparison of functions. Table 10-1a lists the output voltage level as a function of input voltage levels for the ECL gate circuit shown. Table 10-1b translates the voltage levels into the appropriate negative logic levels, which show the function to be $C = \overline{A \cdot B}$; that is, the circuit performs the NAND function. Table 10-1c translates the equivalent positive logic function into $C = \overline{A + B}$, the NOR function.

Similar comparisons could be made for other positive logic functions. As an example, the positive OR function translates to the negative AND function. Figure 10.2 illustrates a comparison of sev-

| POSITIVE LOGIC | NEGATIVE LOGIC |
|---|---|
| **OR/NOR** | |
|  |  |
| **OR/AND, OR/AND/INVERT** | |
| 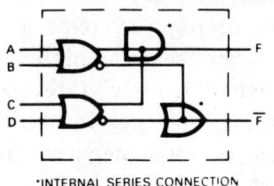 | 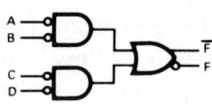 |
| *INTERNAL SERIES CONNECTION NOT THE WIRED OR | |
| **EXCLUSIVE OR/NOR** | |
|  |  |

Figure 10.1   Positive and negative logic.

### TABLE 10.1

#### TRUTH TABLE

| A | B | F | F̄ |
|---|---|---|---|
| L | L | L | H |
| L | H | H | L |
| H | L | H | L |
| H | H | H | L |

(a)

| A | B | C | D | F | F̄ |
|---|---|---|---|---|---|
| L | L | X | X | L | H |
| X | X | L | L | L | H |
| H | X | H | X | H | L |
| H | X | X | H | H | L |
| X | H | H | X | H | L |
| X | H | X | H | H | L |

(b)

| A | B | F | F̄ |
|---|---|---|---|
| L | L | L | H |
| L | H | H | L |
| H | L | H | L |
| H | H | L | H |

(c)

H = Most Positive Voltage — HIGH  
L = Most Negative Voltage — LOW  
X = Don't Care

| INPUTS | | POSITIVE LOGIC | | | | | |
|---|---|---|---|---|---|---|---|
| A | B | AND | OR | NAND | NOR | EXOR | COIN* |
| LO | LO | LO | LO | HI | HI | LO | HI |
| LO | HI | LO | HI | HI | LO | HI | LO |
| HI | LO | LO | HI | HI | LO | HI | LO |
| HI | HI | HI | HI | LO | LO | LO | HI |
| A | B | OR | AND | NOR | NAND | COIN* | EXOR |
| INPUTS | | NEGATIVE LOGIC | | | | | |

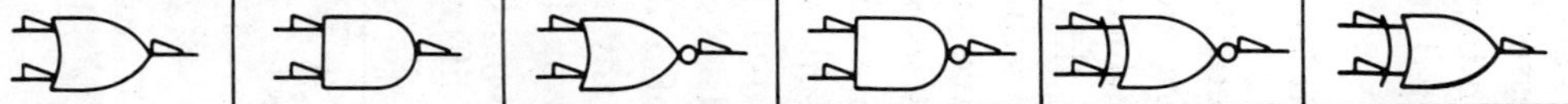

*Coincidence

Figure 10.2 Comparison of logic functions.

**130    Digital Circuit Design**

eral common logic functions. Any function available in a logic family may be expressed in terms of positive or negative logic, bearing in mind the definition of logic levels. The choice of logic definition, as previously stated, is dependent on the designer.

## Noise Immunity

Let us refer to the gate transfer characteristic of Fig. 10.3. It is shown that, for input voltage levels above 2.5 V, the output level is about 0.5 V, which is in the low voltage category. Assume now that the output of this gate is used as the input to a second gate. The output of the second gate would be high because the input was low (you must have guessed by now that we are talking about an inverter). If this second gate were in a system in which electrical interference or noise produced voltage spikes on the input line, the amplitude of such spikes would have to exceed 1.5 V (sufficient to raise the voltage level from 0.5 to above 2 V) if any effect were to be observed on the output. A similar condition would prevail if the normal high output level of 5 V of a gate were used as the input to another gate. In this case, any negative-going voltage spikes at the input of the second gate would have to exceed 2.5 V in amplitude (sufficient to lower the voltage level below 2.5 V) before the output voltage would change. The difference between low and high output voltages of a

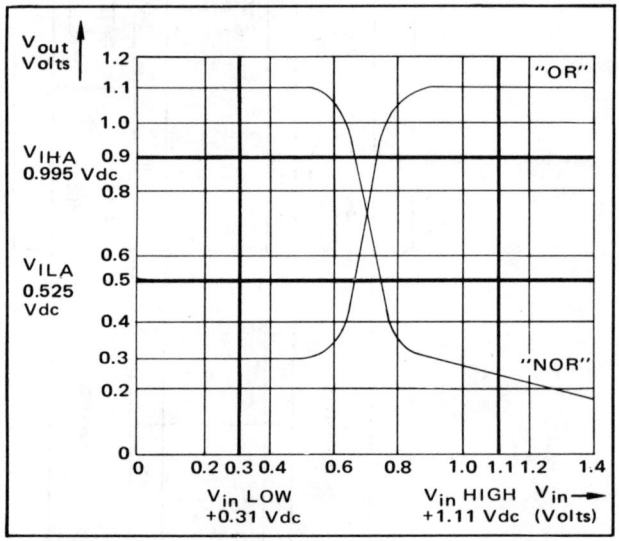

Figure 10.3   ECL transfer characteristics.

gate, and the levels at which the voltage category changes from low to high, and vice versa, is a measure of the noise immunity or noise margin of the gate.

The noise-margin parameter is related to the threshold-level limits. The low noise margin is the difference between the maximum low output voltage level and the minimum low input threshold level. The high noise margin is the difference between the minimum high output voltage level and the maximum high input threshold level. If only one figure is quoted for noise margin, it will be the smaller of the two.

## Threshold Levels

These are the input voltages at which the output voltage begins to change from one level to another. Gates normally have high and low threshold levels. Threshold levels vary for individual gates in a production run, but manufacturers quote a minimum high threshold level and a maximum low threshold level for an integrated circuit.

## Propagation Delay Time

Each gate actually has two propagation delay times, the rise and the fall time. The overall propagation delay time is normally the average of these two figures. Figure 10.4 shows the relationship between the input and output pulses, and defines the rise propagation delay, $t_r$, and fall propagation time, $t_f$. The voltage level, $V_x$, is halfway between the high and low threshold voltage levels.

In practical systems, short propagation delay times are often unwanted. If gates with a relatively slow response are used, many noise spikes that could be present will be so short that the gate circuit will not be able to respond, even though the amplitude of the spikes exceeds the noise margin. Gates with long propagation delay times also are much less critical with regard to the physical layout of a system.

## 10.3 Transistor–Transistor Logic (TTL)

Transistor–transistor logic is often considered the successor of diode–transistor logic (DTL), with the exception that the input diodes in the latter have been replaced by the emitter–base junction of a multiple-emitter transistor.

Let us consider the NAND circuit of Fig. 10.5. This circuit actually consists of five subcircuits, each serving a separate function. The input circuit is a multiemitter transistor functioning as an AND

**132   Digital Circuit Design**

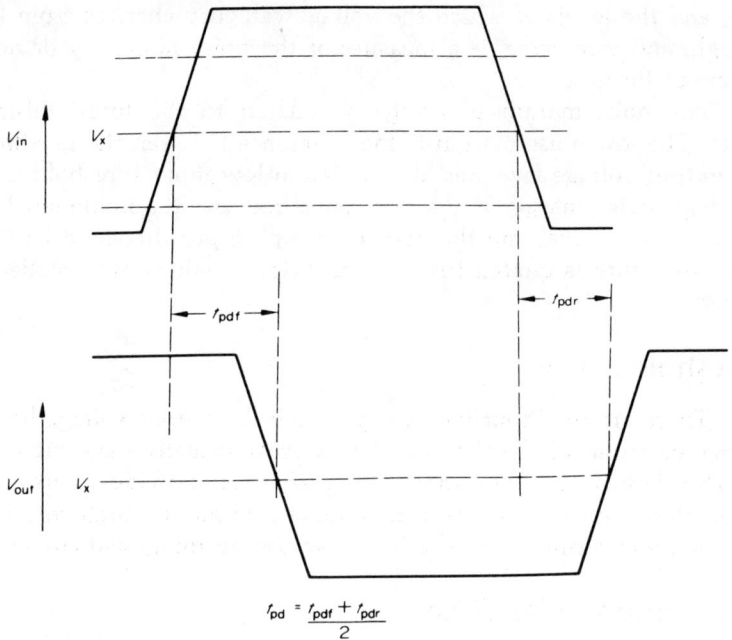

Figure 10.4   Propagation delay time.

gate. When all inputs are high (<2 V), the current in $R_1$ (about 1.2 mA) flows through the collector of $Q_1$ into the base of $Q_2$, turning on $Q_2$, which then turns on $Q_5$, generating a low output voltage. The base of $Q_1$ is clamped three diode drops ($2V_{BE} + V_{CB}$) above ground, about 2 V at room temperature. If any input goes low (<0.8 V), the current in $R_1$ flows through the emitter of $Q_1$, out of the input lead into ground. Any stored charge in the base of $Q_2$ is quickly removed through the transistor action of $Q_1$, and $Q_2$ is thus turned off rapidly. This turns off $Q_5$ and turns on $Q_3$ and $Q_4$, resulting in a high output voltage.

The multiemitter input transistor $Q_1$ has an input leakage current significantly higher than that of DTL. In TTL design, this input leakage current is limited by reducing the inverse current gain of the input transistor through proper choice of transistor geometry and processing. The phase-splitter transistor $Q_2$ produces complementary drive signals for $Q_3$ and $Q_5$. The collector and emitter of the phase splitter can be connected to additional phase-splitting transistors driven by AND gates, thus providing the AND/OR/INVERT function. This configuration is a very useful tool, providing complex logic

## Transistor-Transistor-Logic Outputs

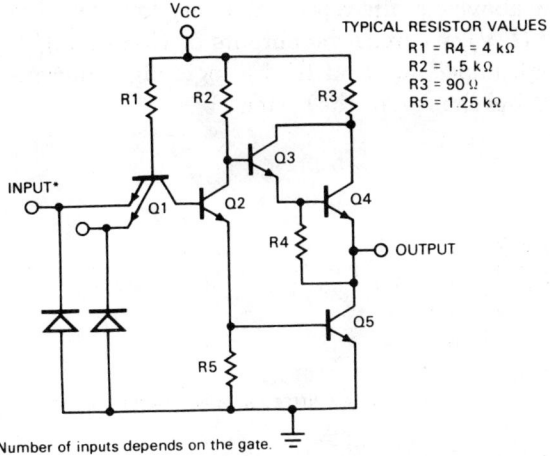

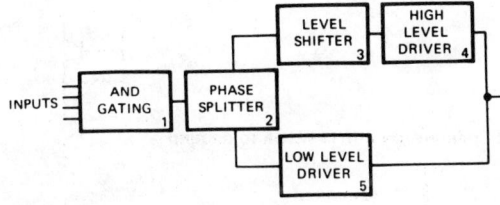

Figure 10.5 TTL NAND gate.

functions with minimum chip area and a minimum number of gate delays.

The Darlington connection $Q_3$–$Q_4$ acts as a low-impedance high-level driver, speeding up the low to high output transition, and providing superior ac drive capability and good noise immunity in the high state. Resistor $R_4$ acts as a current limiter for ac switching currents and for accidental output short circuits to ground. Resistor $R_5$ provides a base turn-off current path for $Q_4$. Connecting $R_5$ to the output rather than to ground reduces power consumption. $Q_5$ is the low-level driver, and $R_5$ provides base turn-off current path for $Q_4$.

## 10.4 Transistor–Transistor-Logic Outputs

Figure 10.6 illustrates the various TTL output configurations, listing their advantages and disadvantages. Circuits (a) and (b) have a diode-resistor combination that clamps the output one diode drop above $V_{cc}$. The importance of this circuit is appreciable in large systems where sections might be powered down ($V_{cc} = 0$). In this state, the outputs of both the above circuits represent a very low impedance at a fairly low voltage (1 V), whereas the outputs of circuits (c), (d), and (e) represent a high impedance and thus a logic high, more appropriate for isolation from the rest of the system.

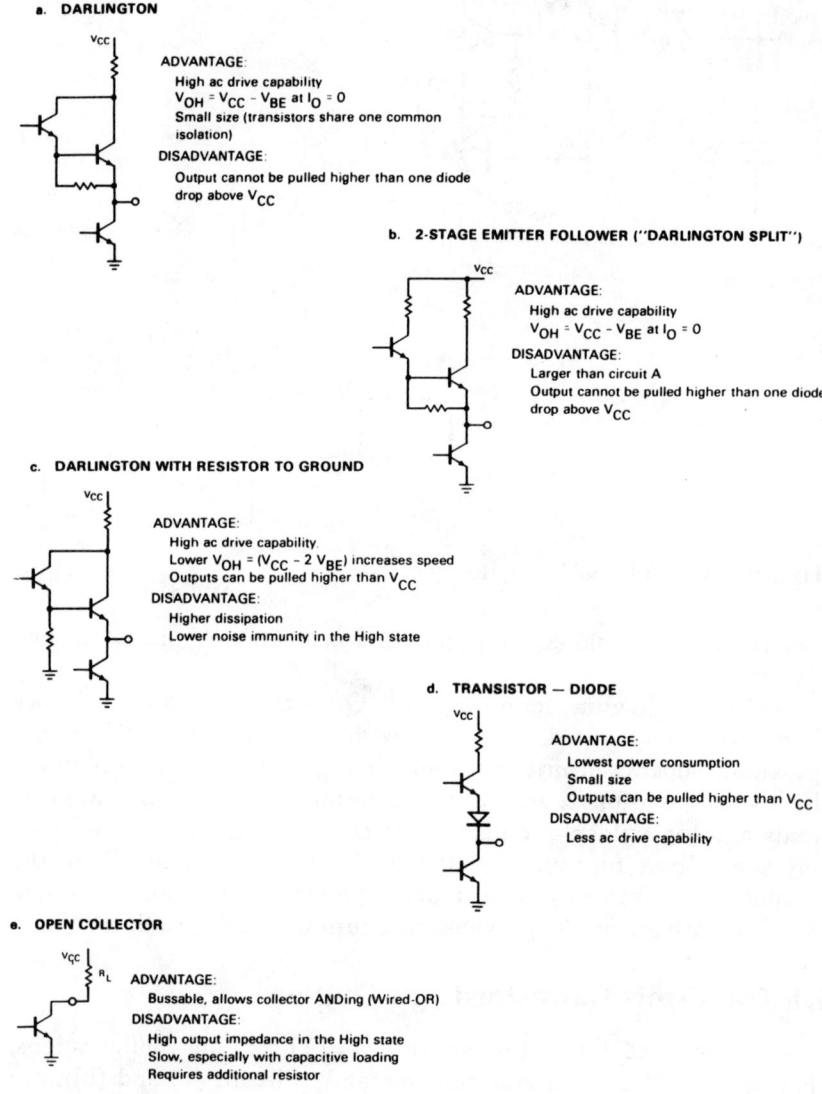

Figure 10.6  TTL output configurations.

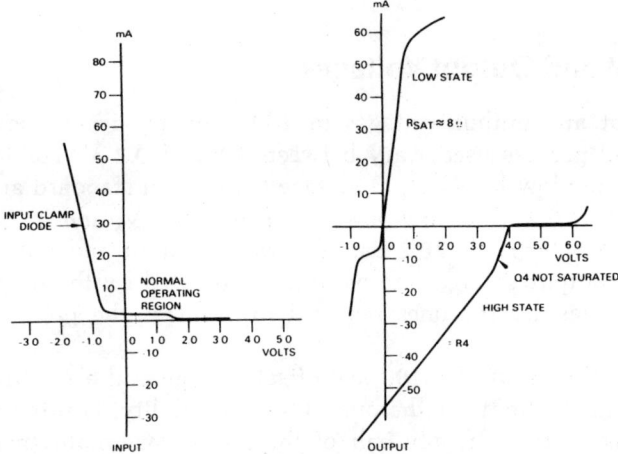

Figure 10.7  Output impedance of TTL device.

## 10.5 Input and Output Impedance

The input impedance of a TTL circuit is very high for positive input voltages (larger than 2 V), about 4 kΩ for voltage between +1 and −0.5 V, and very low for more negative voltages. Each TTL input has a clamp diode to ground that ensures this low input impedance for negative voltages. Clamp diodes protect the input, and limit any negative input swing due to inductive ringing or reflections on transmission-line interconnections.

Figure 10.7 illustrates the output impedance of a TTL device in both the low and high state. In the low state, the output impedance is determined by a saturated transistor (about 8 Ω). However, at very high sinking current, especially at low temperature, the output device is not able to stay in saturation, and the output impedance rises as shown.

When switching from the low to the high state, the totem pole output structure provides a low output impedance capable of rapidly changing capacitive loads. However, charge and discharge currents must also flow through $V_{cc}$ and ground distribution networks. The $V_{cc}$ and ground lines should therefore be short and adequately decoupled. Moreover, if during the low-to-high transition, $Q_5$ has not turned off by the time $Q_4$ is turned on, there is a narrow current spike through the totem pole, which acts as a noise generator unless the supply is properly decoupled.

## 10.6 Input and Output Voltages

Input and output voltages in TTL circuits, depending on the design and process used, vary between 0.1 and 0.2 V. For instance, the maximum low-level output voltage ($V_{OLmax}$) in standard and high-speed TTL is 0.4 V, in low-power TTL it is 0.3 V, and in Schottky TTL it is 0.5 V. Table 10.2 lists the various input and output voltages. The output voltages are naturally generated by the output; the input voltages are the ones required to generate the proper output levels.

$V_{OL}$ is the sum of <100 mV offset voltage and a resistive drop that, within the limits of the output sink capability, is proportionate to the sink current. If only half of the maximum fanout (maximum number of gate inputs to which another gate output can be connected without changing any specified limits and parameters) is used, this voltage for a standard gate will be below 250 mV. The inherent offset voltage for Schottky outputs is higher (about 200 mV).

$V_{OHmin}$ depends on the output configuration. The Darlington approach with a 1-k$\Omega$ resistor to output has the following characteristics:

1. When sourcing negligible current (i.e., under normal stationary operating condition or idling), $Q_4$ is not conducting. Thus $V_{OHmin}$ is one diode drop below $V_{CC}$ with a temperature coefficient of $-2$ mV/°C and an impedance of 1 k$\Omega$.

**TABLE 10.2**

|  | STANDARD AND HIGH-SPEED TTL | LOW-POWER TTL | SCHOTTKY TTL |
|---|---|---|---|
| Maximum low-level output voltage, $V_{OLmax}$ | 0.4 | 0.3 | 0.5 |
| Minimum high-level output voltage, $V_{OHmin}$ | 2.4 | 2.4 | 2.5 |
| Maximum low-level input voltage, $V_{ILmax}$ | 0.8 | 0.7 | 0.8 |
| Minimum high-level input voltage, $V_{IHmin}$ | 2.0 | 2.0 | 2.0 |

2. When sourcing moderate current of a few milliamperes, the $V_{OHmin}$ is two diode drops below $V_{CC}$ with a temperature coefficient of $-4$ mV/°C and a very low impedance (less than 30 Ω).

3. When sourcing large current (20 mA), the Darlington output is saturated, and it follows the resistive load line determined by the current-limiting resistor $R_4$. $V_{IL}$ and $V_{IH}$ have a temperature coefficient of $-4$ mV/°C.

## 10.7 Input and Output Currents

A TTL device has four different currents, input low and high current ($I_{IL}$, $I_{IH}$), and output low and high ($I_{OL}$, $I_{OH}$). To simplify interconnection rules, these currents have been normalized as TTL unit loads (UL). One UL is equivalent to the worst-case input current of a standard TTL, 1.6 mA in the low state and 40 μA in the high state. Input requirements for TTL devices are given in unit loads.

## 10.8 Low Power and Schottky TTL

Low-power TTL offers certain advantages to the designer over standard TTL. This type of logic has derived from medium-scale and small-scale integration of standard TTL by effectively increasing all resistors by a factor of 4, thus reducing all current and power consumption. However, the drawback of low power TTL is that it is slower than standard devices. Thus the designer has the option of using standard devices, where speed is important, and low-power devices where consumption is a significant factor. By the use of low-power devices, power consumption may be reduced by as much as 75 percent, with a consequent reduction in the size of power supplies, heat generated, and noise generated.

Schottky TTL is a satisfactory medium between standard TTL and emitter-coupled logic (ECL), in cases where the speed of conventional TTL is not adequate and the speed of ECL is excessive. A Schottky structure has been discussed previously; Fig. 10.8 illustrates a two-input NAND gate designed with Schottky devices. An advantage of the Schottky circuit is that the saturation delay inherent in saturated circuits is avoided. In saturated logic, transistors are turned on by applying sufficient base current for the lowest expected current gain. Thus the average transistor receives far more base current than necessary, which forward biases the collector–base junction and saturates the transistor. To turn off such a transistor, the excess base charge must first be removed, and this results in considerable

# 138 Digital Circuit Design

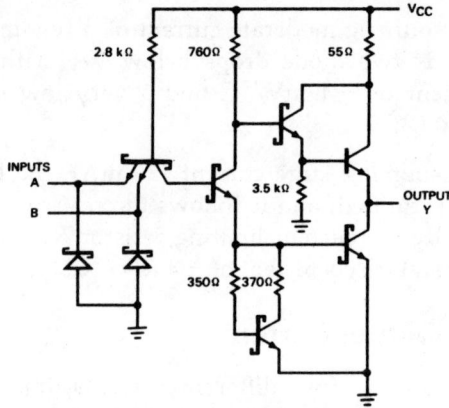

Figure 10.8 Two-input Schottky NAND gates.

delay. One remedy is to use gold doping to speed up the charge recombination, but this decreases the gain.

Schottky transistors overcome this problem by employing a surface barrier diode with very low forward voltage drop (0.3 V) and without minority-carrier charge storage as a bypass between base and collector. When the transistor starts conducting and is about to reach its saturation point, the surplus input current is not fed into the base, but passes through the Schottky diode into the collector. Thus the transistor never reaches a full saturation condition, and, when its base current is interrupted, its recovery is fast. Since gold doping is not necessary, Schottky transistors provide a high current gain, require less base current, and turn on faster.

Thus Schottky devices offer shorter delays, slower increase of static power consumption, and more efficiency. At the same time, they present certain disadvantages over standard TTL. Although Schottky diodes are more efficient and suppress ringing better than standard TTL clamping diodes, they possess a high edge rate, which, even on moderately long (5–10 in.) printed-board interconnections, causes ringing and other transmission-line effects. Schottky interconnections are difficult to terminate without wasting fan-out capability and power. Finally, because the Schottky transistor does not saturate, the maximum output voltage in the low state is 0.5 V, and the noise immunity is about 100 mV less than that of standard TTL.

## 10.9 Emitter-Coupled Logic (ECL)

Emitter-coupled logic offers advantages and capabilities to the designer of high-speed digital systems. High switching rates at relatively low power consumption, short propagation delays with slow edge rates, and the ability to drive low-impedance interconnections are some of the ECL features.

Figure 10.9. illustrates the basic ECL switch. Its logic operation consists of steering the current through either of two return paths to $V_{CC}$; the state of the switch can be detected from the resultant voltage drop across $R_1$ or $R_2$. The net voltage swing is determined by the value of the resistors and the magnitude of the current. These two values are chosen to accomplish the charging and discharging of all the parasitic capacitances at the desired switching rate.

The basic operation of the differential amplifier (switch) is as follows: at idling (both base voltages are equal), the current is equally divided through $Q_1$ and $Q_2$. However, if $V_{in}$ increases by, say, 125 mV, all the current flows through $Q_1$; conversely, if the same voltage is decreased by 124 mV, all the current will flow through $Q_2$. Thus the minimum signal swing required to accomplish switching is 250 mV, centered about $V_{BB}$. In practice, the swing is made large (about 850 mV) to provide sufficient noise margin and to allow for differences between the $V_{BB}$ of one circuit and the output voltage levels of the other.

If the voltage at the collector of $Q_1$ is monitored while varying $V_{in}$ above and below the $V_{BB}$ value, the relationship between $V_{C1}$ and $V_{in}$ is shown as in Fig. 10.10. The horizontal axis of the graph is centered at $V_{BB}$, emphasizing the importance of $V_{BB}$ in establishing the location of the transition region. The shape of the transition region is governed by the transistor characteristics and the value of the current to be switched, and these two factors lie in the hands of the designer.

In Fig. 10.10, $V_{C1}$ ranges from $V_{CC}$ (ground) when $Q_1$ is off to approximately $-0.98$ V when $Q_1$ is conducting all the source current. To make these voltage levels compatible with the voltages required to

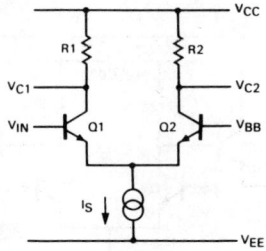

Figure 10.9   ECL switch.

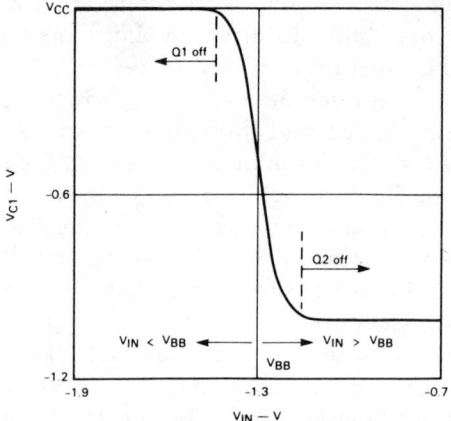

Figure 10.10

drive the input of another current switch, emitter followers are added, as shown in Fig. 10.11. In addition to translating $V_{C1}$ and $V_{C2}$ downward, the emitter followers also isolate the collector nodes from load capacitance and provide current gain. Since the output impedance of the emitter followers is fairly low (about 7 Ω), ECL circuits can drive transmission lines (coaxial cables, twisted pair, and etched circuit) having characteristic impedances of 50 Ω or less.

In the buffered current switch of Fig. 10.11, the collectors of $Q_3$ and $Q_4$ return to a separate $V_{CC}$ lead. This separation ensures that any changes in load currents during switching do not cause a change in $V_{CC2}$ through the small but finite inductance of the $V_{CC1}$ bond wire and package lead. Outside the package, the two $V_{CC}$ leads are normally tied to the common $V_{CC}$ source.

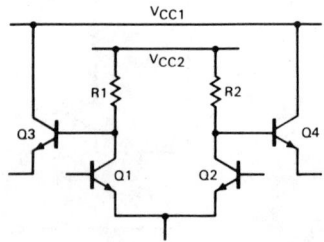

Figure 10.11  Buffered current switch.

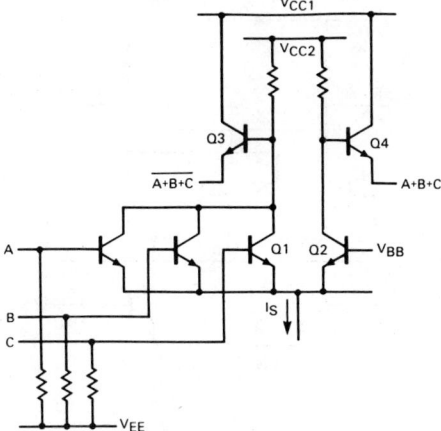

Figure 10.12  Improved ECL switch.

## 10.10  Additional ECL Configurations

The basic ECL switch of Fig. 10.9 can be further modified to perform additional logic functions. For example, to perform the OR and NOR of two or more functions, additional transistors are connected in parallel with $Q_1$, as shown in Fig. 10.12. When any input is high, its associated transistor conducts the source current, and $Q_2$ is turned off; this causes the collector of $Q_1$ to go low and the collector of $Q_2$ to go high, with the emitters of $Q_3$ and $Q_4$ following suit. When two or more inputs are high, the results are the same. Thus, with a level defined as high (logic 1), $Q_3$ provides the NOR function of the inputs while $Q_4$ simultaneously provides the OR function. The resistors in the circuit are bleeding resistors, holding any unused

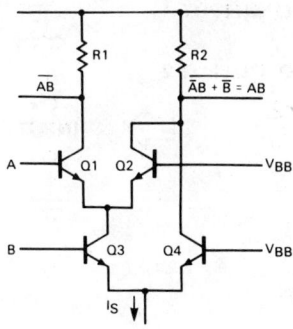

Figure 10.13

**142    Digital Circuit Design**

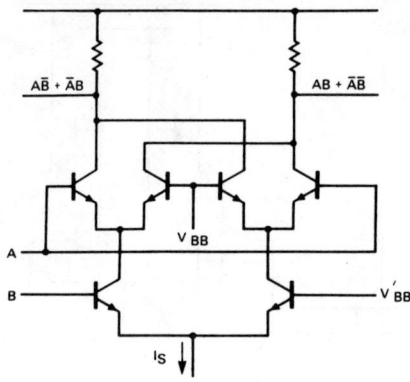

Figure 10.14

inputs to the low state by sinking $I_{CBO}$ current and preventing the buildup of charge on input capacitances.

In Fig. 10.13, if both A and B are high, then $Q_1$ and $Q_3$ conduct, and $I_S$ flows through $R_1$, making the collector of $Q_1$ go low, thereby accomplishing the NAND of A and B. Connecting the collectors of $Q_2$ and $Q_4$ to the same load resistor provides the AND of A and B. If the collectors of $Q_3$ and $Q_4$ were interchanged, a different pair of functions of A and B could be achieved. Similarly, a third functional pair is accomplished by interchanging the collectors of $Q_1$ and $Q_2$.

In Fig. 10.14, by the addition of another pair of transistors over $Q_4$, the OR and Exclusive-NOR function is achieved.

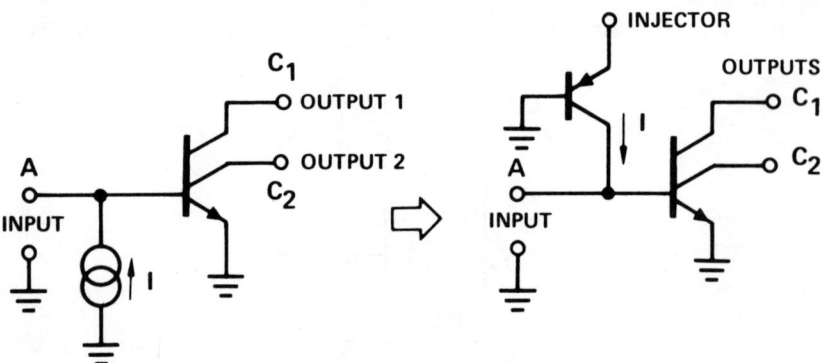

Figure 10.15

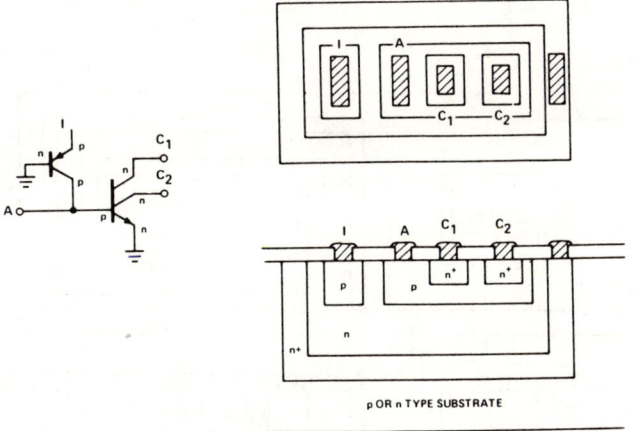

Figure 10.16

## 10.11 Integrated-Injection Logic (I²L)

A rather recent development in the logic family, I²L undoubtedly offers several advantages over TTL and ECL and, although it may not be as fast as they are, it is easier and cheaper to fabricate. The basic building gate of I²L is shown in Fig. 10.15. It is a single-input, multiple-output inverter.

Most terminals of the I²L gate share the same semiconductor region (e.g., the collector of the *pnp* is the same as the base of the *npn*, and the emitter of the *npn* is the same as the base of the *pnp*). This leads to a very compact device structure, and results in very high packing density in monolithic device fabrication. Figure 10.16 shows the basic device cross section for a two-output gate. This basic structure can be made compatible with basic bipolar IC technology by using a silicon *p*-type substrate as the semiconductor starting material. The bipolar compatible I²L device structure is shown in Fig. 10.17. Figure 10.18 shows a size comparison of the basic four-output I²L gate, with a four-input TTL gate, each fabricated with the same masking tolerances. Note that I²L offers a 5-to-1 reduction in gate area.

## 10.12 Advanced Logic Circuits

The success of any logic family depends on the capability of its storage elements. Flip flops are logic circuits used in frequency-divi-

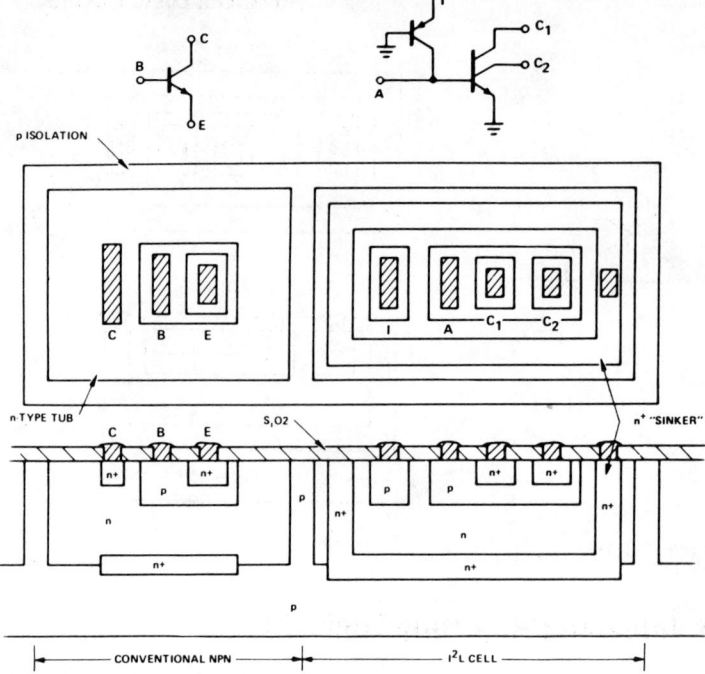

Figure 10.17

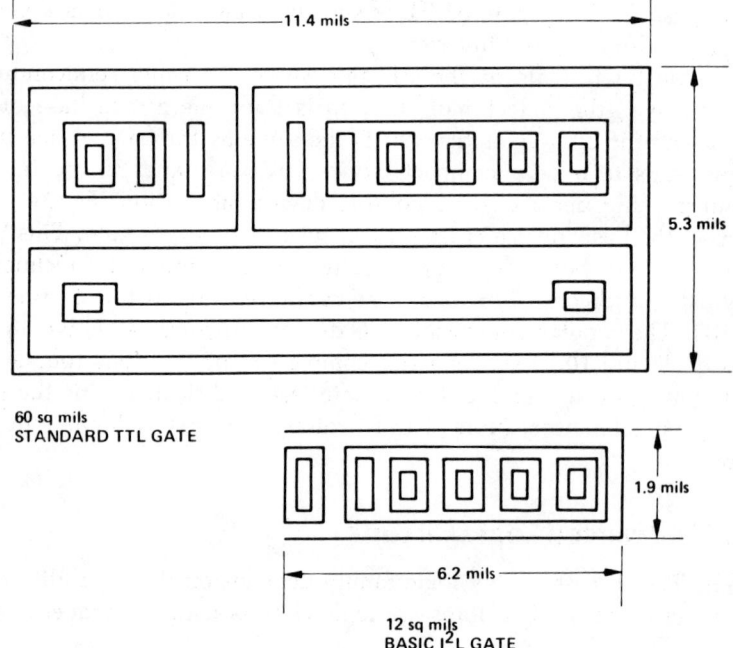

Figure 10.18

Advanced Logic Circuits   145

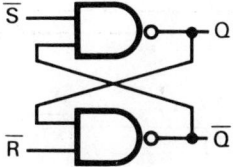

Figure 10.19   R-S flip flop.

sion counting, special coded counters, shift registers, and storage registers. They offer high-speed operation and logic versatility.

One type of flip flop is the set–reset (R–S) flip flop. It is a bistable logic element in which one or more inputs will set the device to the logic 1 state (high), and one or more inputs will reset the device to the logic 0 (low) state. The R–S flip flop, as shown in Fig. 10.19, is similar to two NAND gates with the outputs cross-coupled back to the inputs.

A more sophisticated version of the R–S flip flop is shown in Fig. 10.20. The main difference between the two circuits is that the latter version has an additional buffering on the $Q$ and $\bar{Q}$ outputs which prevents the flip flop from changing state in high noise level output environments.

Another type of flip flop is the $\bar{J}$–$\bar{K}$ flip flop. The advantage of this circuit is that no undefined output conditions can exist. A circuit of this type is shown in Fig. 10.21. A relatively high level voltage ($-0.75$ V) is defined as logic 1, while the logic 0 corresponds to $-1.55$ V. The $\bar{J}$ and $\bar{K}$ inputs represent static logic levels; $\bar{C}_D$ refers to a dynamic logic swing. The $\bar{J}$ or $\bar{K}$ inputs should be changed to a logic 1 while $\bar{C}_D$ is at a high level; otherwise, the $\bar{J}$ or $\bar{K}$ signal may clock the flip flop. The set–reset operation is the same as that of the R–S flip flop.

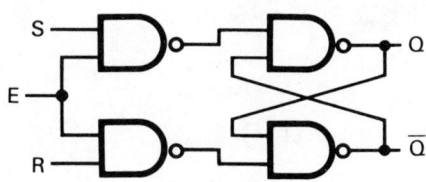

Figure 10.20   Improved R-S flip flop.

**146    Digital Circuit Design**

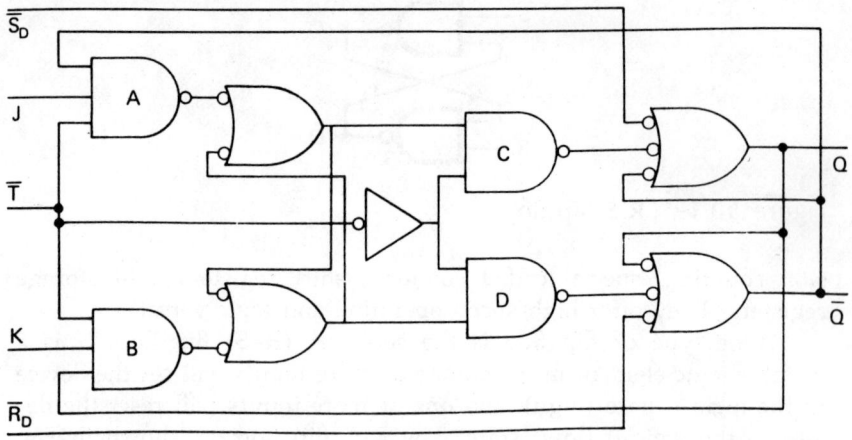

Figure 10.21   J-K flip flop.

The discussion of the digital circuits in this chapter provides a complete description of the various design techniques, and is a complete guide for more advanced designs.

## Exercises

**10.1** Describe the various advantages and disadvantages of the logic families.

**10.2** Describe noise margin and provide a graphical example.

**10.3** An OR gate is connected to the inputs of four AND gates. What is the fan out of the OR gate?

**10.4** The inputs A, B, and C to a logic system change regularly with time, in accordance with the following table. The output of the system is required to be low if $T = n, n + 1$, or $n + 7$ ($n = 1, 2, 3$, etc.), and high otherwise. Show how the system could be implemented using 2 three-input AND gates, 1 two-input OR gate, and 1 R–S flip flop.

|      | Inputs | | |
|------|---|---|---|
| Time | A | B | C |
| 0 | 0 | 0 | 0 |
| 1 | 0 | 0 | 1 |
| 2 | 0 | 1 | 0 |
| 3 | 0 | 1 | 1 |
| 4 | 1 | 0 | 0 |
| 5 | 1 | 0 | 1 |
| 6 | 1 | 1 | 0 |
| 7 | 1 | 1 | 1 |
| 8 | 0 | 0 | 0 |
| 9 | 0 | 0 | 1 |
| etc. | | | |

**10.5** Condition requirements: short delay, low power, and 12-in. lead length between device and output terminals on a board. Which type of logic family would you recommend for this system?

**10.6** What is the difference between DTL and TTL?

## chapter eleven
# MOS and CMOS Circuit Design

## 11.1 Introduction

As stated previously, the heart of the MOS (metal-oxide semiconductor or metal-insulator semiconductor, MIS) is the MOSFET. This voltage-controlled device maintains a high input resistance that is independent of the magnitude or polarity of the input gate voltage. MOS integrated circuits have become an extremely vital device in the electronic industry with the development of microprocessors and memory devices. CMOS denotes complementary symmetry MOS. For the reader to fully understand the MOS design theory, certain portions from previous chapters are summarized here.

## 11.2 Basic Operation of the MOS Transistor

The symbol of a MOSFET is shown in Fig. 11.1. When both the drain and the source are grounded, the gate controls the charge in the channel (channel is the area of the substrate surface between the source and the drain). When the gate is supplied with a negative voltage, free electrons in the $n$-type silicon are repelled, thus forming a depletion region. When depletion reaches completion, the additional gate bias begins to attract positive holes to the surface. When suffi-

## Basic Operation of the MOS Transistor

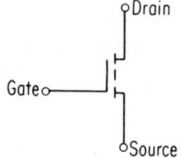

Figure 11.1  MOSFET.

cient holes have accumulated in the channel area, the surface of the silicon inverts, that is, changes from a predominantly negative material to a predominantly positive material. Thus the two *p*-type diffused regions are connected together by a *p*-type inversion layer (*p*-channel device). Then, from this action, it is evident that the gate controls the current flow in the channel.

When the gate is fed with a constant voltage, the situation changes by increase in the drain voltage. The drain current produces an *IR* drop along the channel, and this drop is of such polarity that it opposes the field within the oxide produced by the gate bias. When this *IR* drop reaches a sufficient value to reduce the field and to overcome the inversion layer, the channel pinches off, and the drain current tends to saturate at a constant value. Thus the transistor is saturated. The voltage across the gate oxide, just at the point of saturation, is called the *pinch-off* voltage or threshold voltage, and, in actual application, it is the voltage across the gate oxide required to just produce channel inversion. Any further increase in drain voltage will drive the transistor further into saturation, and will eventually cause the drain depletion region to punch through all the way to the source, and to generate unlimited current. External circuitry most of the time prevents this action.

The main differences between an MOS and a bipolar device are the following:

1. The bipolar device is controlled by a current, whereas the MOS is controlled by a voltage.

2. Input–output voltages and currents all have the same sign (a very convenient quality for cascading stages in digital circuitry). It is the opposite case with a junction FET.

Although the MOS device somewhat lacks speed, compared to bipolar devices, it offers several advantages, as follows:

1. Parasitic resistances are so small that they are negligible.

**150**  MOS and CMOS Circuit Design

2. The channel is completely shielded from the drain, so no drain-to-channel feedback exists.

3. Doping of the substrate is uniform and nondegenerate.

4. The drain current consists only of channel current, and leakage currents are neglected.

5. The gate dielectric is considered to be a perfect insulator.

A MOSFET can be operated in three modes, depletion, enhancement, and a combination of the two. In the depletion mode, the MOSFET has substantial drain current flow at zero gate voltage. Drain current is reduced by applying a reverse voltage to the gate terminal. In the enhancement mode, the device has negligible or zero current at zero gate bias. Drain current is not noticed until a forward gate voltage is applied (this voltage, as discussed previously is the threshold voltage). Finally, in the combination of the two, a substantial drain current is present, but not as much as in the depletion mode. This current is increased by forward gate voltage and decreased by reverse gate voltage. The $V$–$I$ characteristics of a device are shown in Fig. 11.2, where we can see that, in part a, no significant current flows until the voltage is brought up to $-4$ V (enhancement mode). In part b, a current of $-85$ $\mu$A flows at zero gate bias (depletion mode).

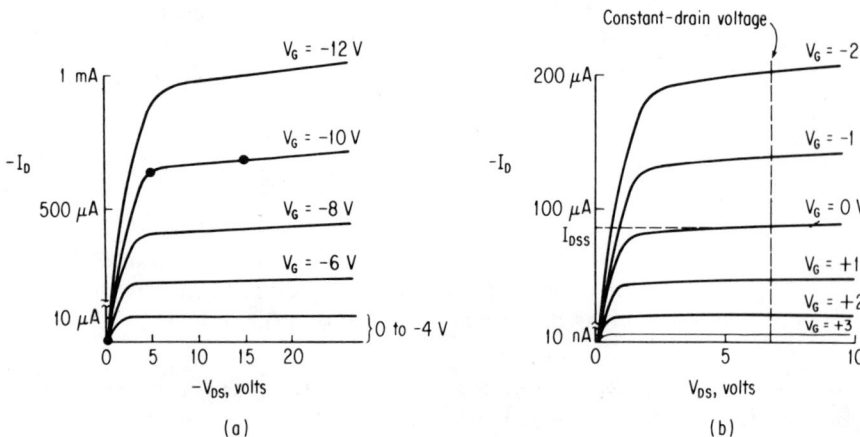

Figure 11.2   MOSFET V-I characteristics.

## Interpretation of Symbols    151

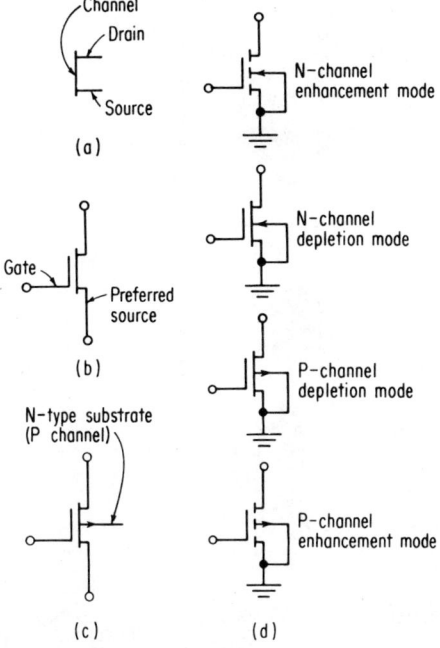

Figure 11.3   MOS symbols.

## 11.3 Interpretation of Symbols

The most commonly used symbols for MOS devices are shown in Fig. 11.3. Their explanation is as follows:

1. The source and drain are ohmic contacts, and thus are shown at right angles to the channel.

2. Enhancement-mode devices do not have current flow for zero gate bias, and thus are represented by a dashed drain line. Depletion-mode devices with initial conduction at zero gate bias are depicted by a solid line.

3. MOS gates, because they are insulated and are not a *pn* junction, are depicted with an L-shaped symbol with one side parallel to the channel, and the corner of the gate is placed opposite to the preferred source.

4. The substrate is shown as a nonemitting diode and, as such, is drawn perpendicular to the channel. The direction of the arrow indicates the type of conductivity of the substrate.

## 152    MOS and CMOS Circuit Design

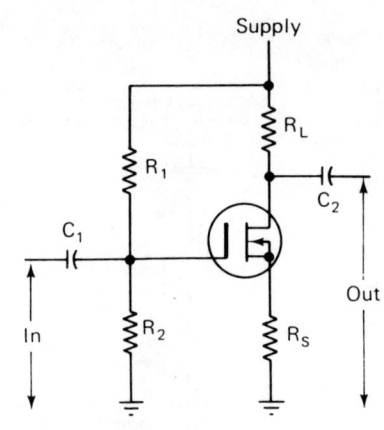

$$A_V \approx \frac{R_L}{\frac{1}{Y_{fs}} + R_S} \approx \frac{R_L}{R_S} \quad \text{Drain voltage} = 0.5 \times \text{supply}$$

$$Z_{IN} \approx R_1 \parallel R_2 \approx R_2$$
$$Z_{OUT} \approx R_L \qquad\qquad \text{Minimum } R_L \approx R_S \times \text{gain}$$

Figure 11.4    Common-source circuit.

## 11.4   Basic MOS Circuits
### Common Source

The common-source configuration is shown in Fig. 11.4. The gain of the stage is given by

$$A_v = -g_m R_L = \frac{R_L}{R_S}$$

where

$g_m$ = transconductance
$R_L$ = load resistance
$R_S$ = source resistance

### Common Drain (Source Follower)

This is a most useful basic circuit, and it is shown in Fig. 11.5. Several of its advantages are that it presents no phase reversal at the output, the voltage gain is always less than unity, its output impedance is low, and it has a large-signal swing.

The gain of the circuit is given by

$$A_v = \frac{g_m}{1 + g_m R_S} R_S$$

If we assume that the transconductance is 1000 $\mu$S, and $R_S$ is 5 k$\Omega$, the source-follower voltage gain is +0.835. Doubling the source resistance will increase the gain to only +0.91.

The output impedance is given by

$$\frac{i_o}{v_o} = G_S + g_{ds} + g_m$$

where

$G_S$ = external source conductance
$g_{ds}$ = saturated drain conductance

Thus the output impedance of the previous example is

$$\frac{5(10)^3}{10^{-3}(5)(10^{+3}) + 1} = 835 \; \Omega$$

Doubling the transconductance will cut the output impedance to almost half, or 445 $\Omega$.

The bias current is the voltage across the source resistance divided by that resistance, or

$$I_D = \frac{V_{R_S}}{R_S}$$

## Common-Gate (Grounded) Configuration

This configuration is shown in Fig. 11.6. Its gain is given by

$$A_v = g_m R_L$$

## 11.5 Integrated Circuits

Integrated MOS circuits offer several advantages over bipolar devices. When examining an MOS IC, one will undoubtedly notice the lack of passive components, such as resistors and capacitors. This simplifies design and cuts down the cost. The reasons are that (1) the MOS effectively performs as a diffused resistor, (2) because of its

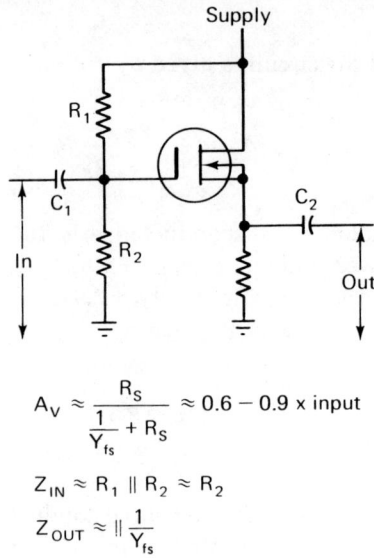

$$A_V \approx \frac{R_S}{\frac{1}{Y_{fs}} + R_S} \approx 0.6 - 0.9 \times \text{input}$$

$$Z_{IN} \approx R_1 \parallel R_2 \approx R_2$$

$$Z_{OUT} \approx \parallel \frac{1}{Y_{fs}}$$

Figure 11.5   Source follower circuit.

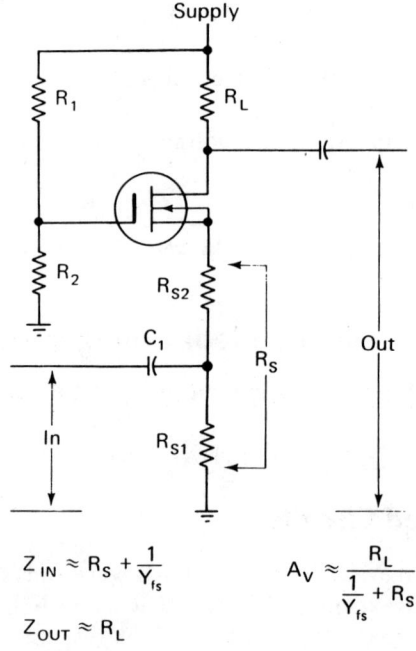

$$Z_{IN} \approx R_S + \frac{1}{Y_{fs}}$$

$$A_V \approx \frac{R_L}{\frac{1}{Y_{fs}} + R_S}$$

$$Z_{OUT} \approx R_L$$

Figure 11.6   Common-gate circuit.

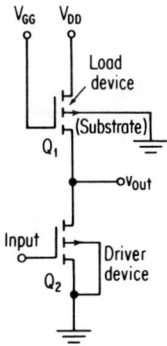

Figure 11.7 MOS inverter.

ability to direct couple, it eliminates the necessity for coupling capacitors, and (3) multiple-clocking schemes, rather common in MOS IC's, eliminate clocking capacitors.

A rather important factor of the MOS process is the high sheet resistivity (20 k$\Omega$/square as compared to 100 $\Omega$/square for bipolar devices). This allows fabrication of more complex devices in a much smaller area.

A basic building block for MOS circuits is the inverter, which is shown in Fig. 11.7, with its equivalent circuit layout in Fig. 11.8. The latter illustration clearly designates the various regions in the layout. The middle portion (combination function of source and drain) is of particular importance, because it reduces the overall chip area during fabrication of MOS IC's. Another space-saving factor results from the elimination of isolation diffusion, owing to the fact that $p+$ regions are inherently isolated from the $n$-type substrate. A further advantage in the layout of Fig. 11.8 is that the load gate is returned

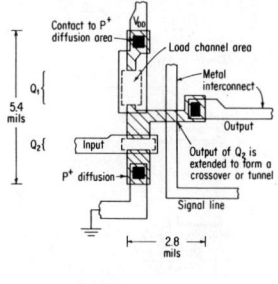

Figure 11.8 Layout of inverter.

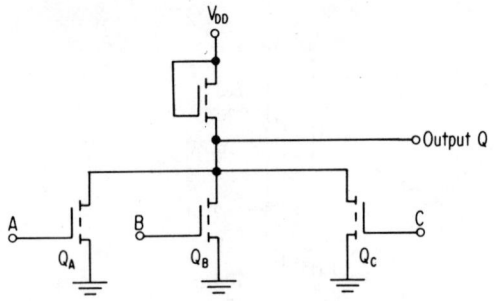

Figure 11.9  Parallel arrangement of MOS devices.

to the load drain and $V_{DD}$, and thus the circuit requires only one power supply and one interconnecting lead.

Figure 11.9 illustrates a parallel arrangement of MOS devices that is rather common in MOS circuitry. The particular circuit under discussion is a three-input NOR gate. From the illustration, it is seen that the three drivers have all the same width-to-length ratio, and thus equal transconductance. The $g_m$ of the transistor $Q_1$ is designed so that $I_D/g_m$ yields the desired $V_{ON}$. Any additional devices paralleling $Q_1$ will not change the $g_m$, but will only tend to lower $V_{ON}$ when more than one driver is turned on by the input logic.

A series arrangement of MOS logic is shown in Fig. 11.10; it is a NAND gate. In this circuit, the NOT function occurs when A and B and C are 1. The ON voltage across the combination of $Q_1$, $Q_2$, and

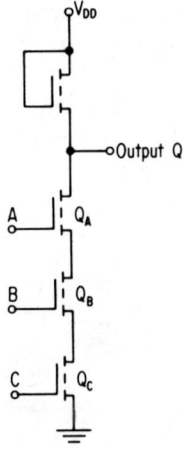

Figure 11.10  Series arrangement.

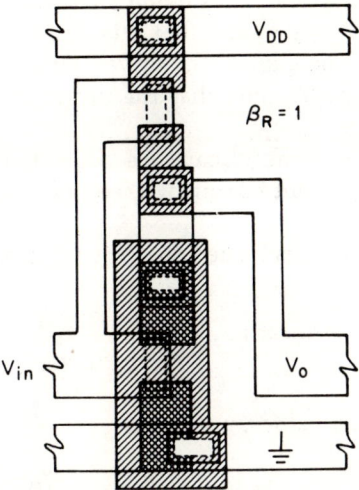

Figure 11.11  Equivalent layout of circuit.

$Q_3$ is the product of the ON current and the total resistance to ground. Since the devices are in series, their resistances add, and thus $r_{ON}$ (total) $=3/g_m$ (for equal geometries). For a given $V_{ON}$, the $g_m$ (and, consequently, the geometry) of the individual devices must be at least three times larger than for the case of a single device. As the number of devices in series increases, so do the $g_m$ and the area. Figure 11.11 is the equivalent layout of the circuit.

The following is an example of the difference in $g_m$ between NAND and NOR circuits. Figure 11.12 illustrates a three-input com-

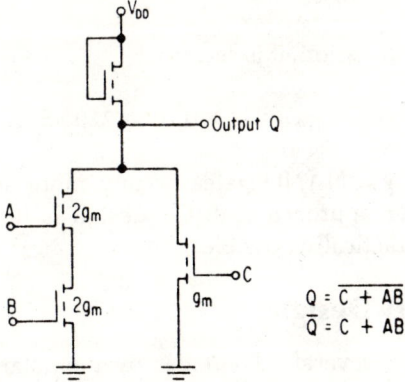

Figure 11.12  Three-input NAND-NOR gate.

bination NAND–NOR gate. Assuming that $V_{DD} = -15$ V, that $V_{th}$ (threshold voltage) $= -4$ V, and that it is desired to let $V_{ON} \leq 0.5$ V, we are asked to find the gain of the driver devices when $g_{m1} = 10$ μS. In this circuit, we also assume that the driver devices are driven by similar inverter stages.

To better analyze the circuit, we draw the equivalent of Fig. 11.13. From the equivalent circuit, we see that $V_{ON}$ must be $\leq 0.5$ V when either A and B or C is turned on. When C is on

$$V_{ON} \cong (V_{DD} - V_{th}) \frac{1}{2} \frac{g_{m1}}{g_{mC}}$$

Substituting numerical values, we obtain

$$-0.5 \text{ V} = (-15 + 4) \frac{1}{2} \frac{10}{g_{mC}}$$

Thus, $g_{mC} = 110$ μS.

Since $V_{ON}$ must be the same, irrespective of which side is conducting, the ON voltage of the left side is equated to that of the right side, and the equation is solved for $g_{mA}$ and $g_{mB}$:

$$V_{ON} = (V_{DD} - V_{th}) \frac{g_{m1}}{g_{mAB}}$$
$$= (V_{DD} - V_{th}) \frac{1}{2} \frac{g_{m1}}{g_{mC}}$$
$$g_{mAB} = 2g_{mC}$$

Thus, from the last equation, it is seen that

$$g_{mA} = g_{mB} = 2 \times 110 = 220 \text{ μS}$$

Due to the higher $g_m$, NAND gates occupy more space than NOR gates. Thus a better approach to MOS design is implementation of NOR gates when practically possible.

## 11.6 MOS Memory Designs

Owing to their several advantages over bipolar systems, MOS devices have found wide application in memory systems. The heart of an MOS memory system is the basic memory cell shown in Fig.

MOS Memory Designs    159

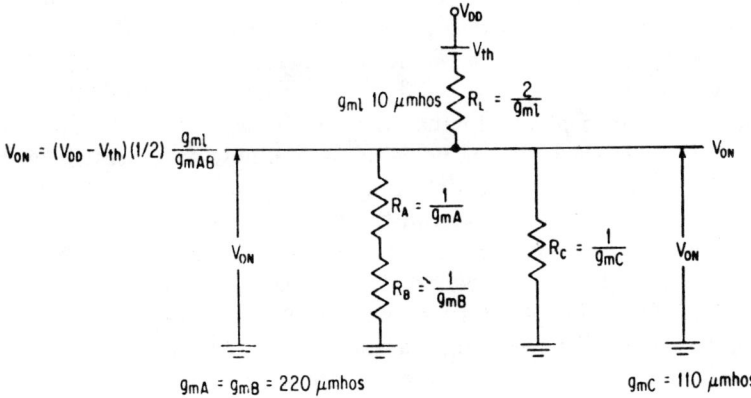

Figure 11.13   Equivalent circuit.

11.14. This type of cell is categorized as *dynamic logic*. It consists of three transistors, and stores data on the gate capacitance C of $Q_2$. Since data storage is accomplished on this gate capacitance, additional circuitry is required to keep these capacitors charged.

A typical memory consists of 1024 bits of storage, organized in a square matrix of 32 rows by 32 columns. During the time interval, which we shall call *precharge*, and when the unit is low, one row is selected by the row decoder. When C enable (chip enable) goes low, all 32 memory cells of the selected row send their data through $Q_3$ to their respective column refresh buffers. The refresh buffers are also connected to 32 column read/write buffers.

When precharge returns high, the data stored in the 32 refresh

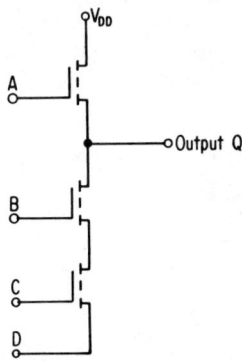

Figure 11.14   Basic memory cell.

buffers are returned to the cells of the selected row via $Q_1$, and the dynamic column address decoder selects one column buffer.

At the end of a data out timing interval, the selected column read/write buffer output is valid. If new data are desired, the read/write amplifier is placed in the write mode at the end of the read/write timing interval. In the write mode, the new data present on the column input override the data previously stored in $Q_2$. At the end of C enable, $Q_1$ is turned off and the data are held on the gate of $Q_2$.

Since the memory cell is a charge-storage type, charge will leak off the gate of $Q_2$ over a period of time, and the stored data will be lost. Therefore, the cell must be periodically recharged to retain the data. To ensure that data are maintained, the cell must be recharged (or refreshed) at least once every 2 msec. To reduce the amount of time spent refreshing, the memory is organized so that all 32 memory cells in a row are refreshed during one read cycle. Therefore, only 32 refresh cycles are required every 2 msec to completely refresh the entire memory. This means that a refresh cycle must be initiated every 62.4 $\mu$sec. This represents less than 1 percent of the total available cycle time when the memory is operated at maximum frequency.

## Exercises

**11.1** Briefly discuss the advantages of MOS technology over bipolar.

**11.2** The given schematic is an R–S flip flop. Analyze the circuit and interpret its symbols. Give a description of the circuit's operation.

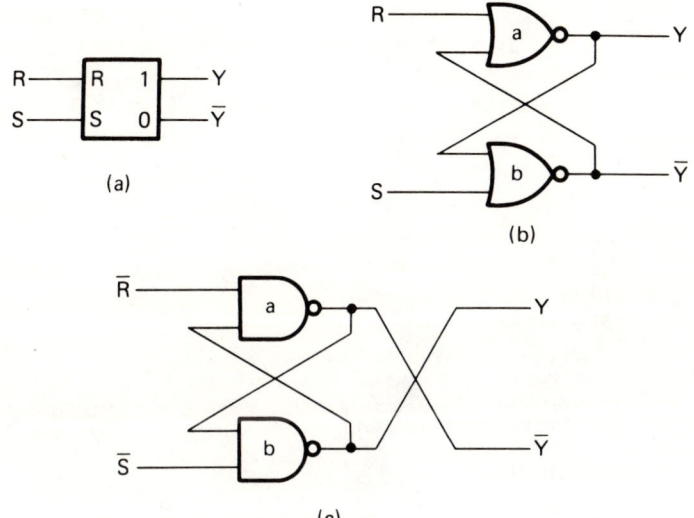

**11.3** Find the total $r_{ON}$ resistance in a five-input NAND gate when the transconductance is 150 µS.

**11.4** Describe the operation of the circuit of Fig. 11.7.

**11.5** Describe the operation of a memory cell.

# chapter twelve
# Large-Scale Integration

## 12.1 Introduction

Most of the circuits discussed so far are frabricated under either small-scale or medium-scale integration techniques and rules. However, the wonderful development of large-scale integration (LSI) has provided the designer with means of producing such complex and large circuits as microprocessors and memories.

The impact of large-scale integration on the economics of the industry is rather evident. Figure 12.1, which compares bipolar and MOS prices over the years, defines just what the new bipolar LSI means in terms of price. The most important point is that, in their new LSI form, bipolar circuits have begun to track MOS prices. Naturally, one of the main reasons that gate prices have been dropping off so quickly is not only the development of LSI, but also the achievement of obtaining LSI at high yields. The higher the yield of a complex circuit, the lower the price will be.

MOS has a parallel history. Once it achieved LSI status, prices plunged, and a 3000-gate chip is now priced at less than $2, or less than .1 cent per gate, while a 4096-bit memory sells for less than $4, or .1 cent per bit.

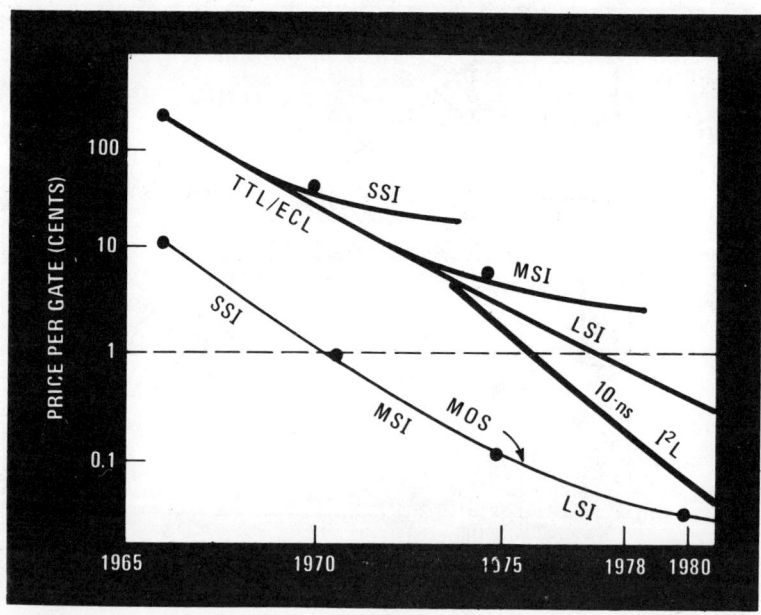

Figure 12.1  Comparison of price per gate

## 12.2 Integrated-Injection Logic (I²L)

Although briefly covered previously, owing to its favorable acceptance, I²L deserves wider discussion in this chapter on LSI to which it most appropriately belongs. Figure 12.2 underlines the various bipolar logic techniques and also places them in contrast with present-day MOS techniques. It is clearly shown that the widest range of applications belongs to I²L, since it can achieve speeds of 10–100 nsec on chips containing 1000 gates or more. Being roughly the application area of present TTL circuits, these figures indicate that present hardwired TTL designs can very well be replaced with microprogrammable I²L designs at much lower cost.

Figure 12.3 illustrates a comparison of the various fabrication techniques in logic. The I²L gate is the smallest that can be built with any of the present circuit techniques, including MOS. It is also the simplest to fabricate, requiring only four masks and two diffusions. For higher performance, another diffusion is required for Schottky structure. Unfortunately, the inverted I²L transistor may not be as easy to fit with Schottky clamps as the standard TTL gate, and in-

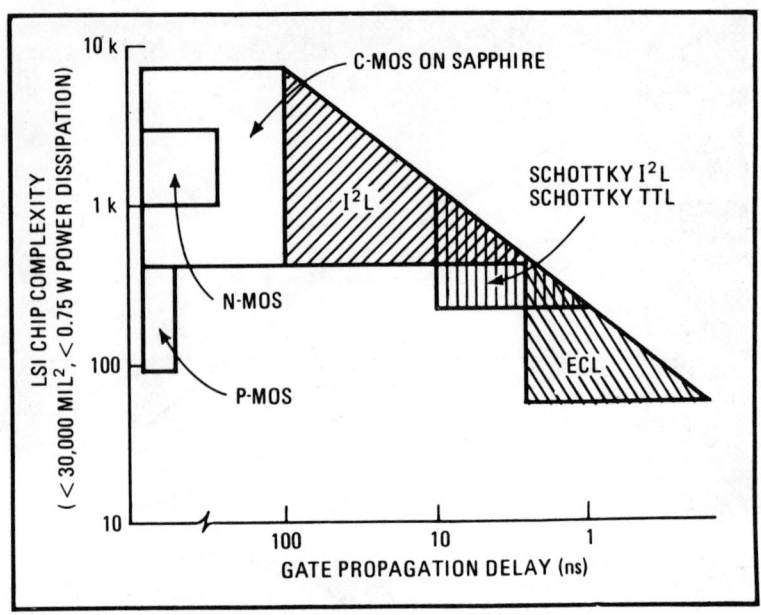

Figure 12.2  Comparison of I²L with other approaches

genious processing may be required. Nevertheless, since the I²L gate is already so small and its power dissipation so low, it might well prove capable of absorbing the added structures and extra processing steps, and still end up smaller than a low-power Schottky TTL gate.

The simplicity of the basic I²L gate is evident from Fig. 12.4, which presents a conventional I²L gate, a Schottky I²L gate, and, most elegant of all, an advanced pure-metal-Schottky design proposed by International Business Machines. In all three cases, the basic logic unit is simply an inverter, implemented as a multicollector transistor or, in other words, a conventional multiemitter transistor operated inversely. Base drive to the *npn* element is supplied from the collector of a lateral *pnp* transistor operating in a current-source mode. The emitter of this lateral *pnp* serves as the gate injector. The basic structure derives much of its appeal from the fact that a multiplicity of *npn* inverters may be powered from a single *npn* emitter or injector, which then distributes current to all the units that form its multicollector.

MOS/CMOS LSI    165

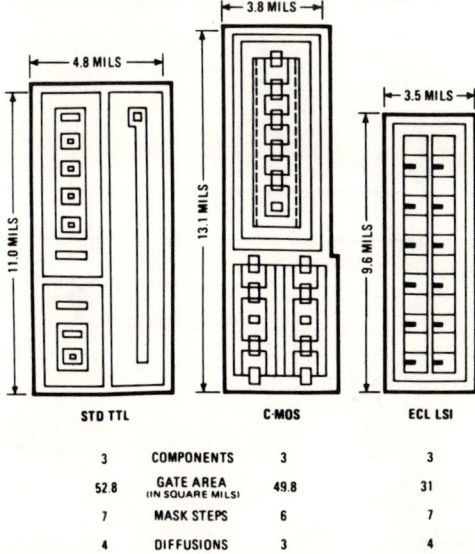

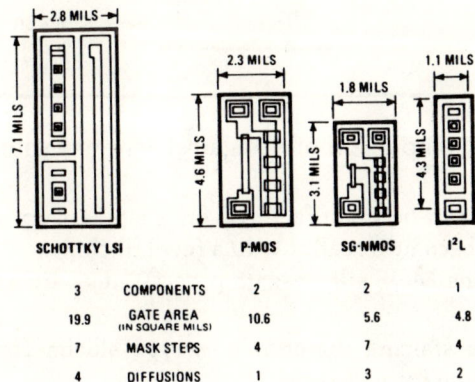

Figure 12.3   Comparison of fabrication techniques

## 12.3 MOS/CMOS LSI

Great advances have been also made in the field of MOS, and the silicon-gate MOS process briefly described next has had a tremendous impact on the design of random logic chips, because it allows the high component and interconnect densities that are required to economically produce these chips. With this technology, a complete cen-

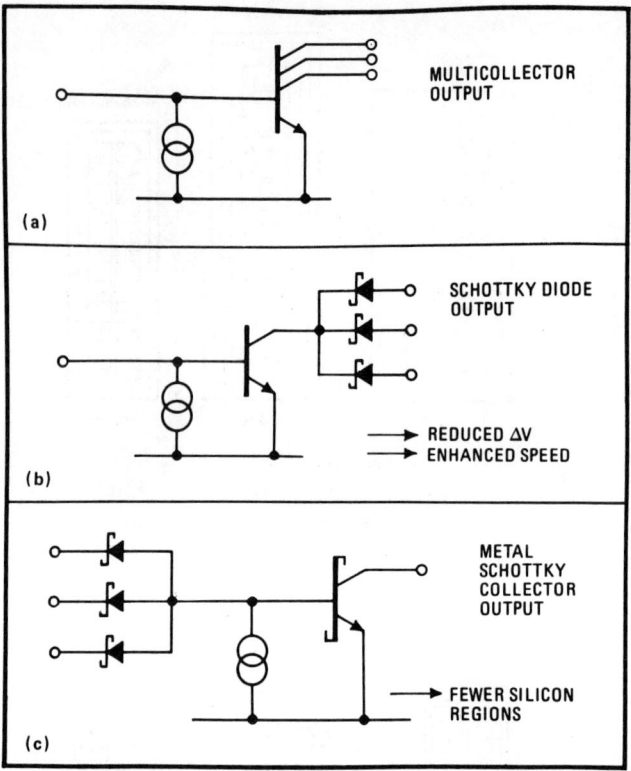

Figure 12.4  Comparison of I²L against Schottky gates

tral processor can be manufactured on a single chip, and a complete computer has become a reality with a few LSI chips.

Briefly, the MOS silicon-gate process steps are as follows:

1. The starting material is $n$-type silicon for $p$ channel or $p$-type for $n$ channel.

2. The wafer is placed in an oxidized atmosphere at high temperature, and a relatively thick layer (about 1 $\mu$m) of SiO₂ is grown on the surface.

3. The region for the source, drain, and channel is defined by photomasking and etching.

4. The wafer is again placed in an oxidizing atmosphere, and a thin layer of SiO₂ (0.1 $\mu$m) is formed. This layer of SiO₂ will serve as the gate dielectric.

5. A thin layer of polycrystalline silicon is deposited over the entire wafer.

6. The wafer is returned to photomasking for removal of the silicon layer except where the gate regions are defined or where the silicon film will be used as an interconnection. The thin oxide is then removed by exposing the wafer to an oxide etch.

7. The wafer is placed in a diffusion furnace where boron (for $p$ channel) or phosphorus (for $n$ channel) is diffused into the gate, interconnect, source, and drain regions.

8. A thick layer of oxide is deposited over the entire wafer, and openings are etched for contacts between the subsequent metallization and underlying diffused regions or the polycrystalline silicon interconnect level.

9. Aluminum is evaporated over the entire surface and is etched to define the metal pattern.

10. The process is completed with the coating of a layer of glass over the entire chip and the etching of the contact holes for the external interconnect.

The silicon-gate process has many inherent advantages over the conventional metal-gate MOS process. These advantages have made possible the fabrication of low-cost random-logic LSI devices. Following are two examples of the contribution of this process, and more particularly, of LSI. One of the first microprocessor chips was the 4004 CPU built by Intel. This device has a chip size of 117 × 159 (approximately 18,500 mils$^2$), which includes 2248 transistors. Another popular CPU has been the 8008, again built by Intel. This device has a chip size of 124 × 173 mils (approximately 21,400 mils$^2$) and includes 3098 transistors.

## 12.4 The (111) Standard Process (P-Channel Enhancement Mode)

The process described in Section 12.3 is by definition the method employed in the fabrication of $p$-channel enhancement-mode MOS/LSI on (111)-oriented silicon. The silicon gate is a (111)-oriented process. To further understand the (111) process (100 and 110 plane process methods will be discussed in later sections), let us consider the following example:

## Large-Scale Integration

In MOS technology, it is frequently important to specify various planes in a silicon crystal. Miller indexes are employed for this purpose, and are specified as follows:

1. Determine the intercepts $a$, $b$, and $c$ of a plane with the three axes $x$, $y$, and $z$, respectively. The plane is chosen in such a way that no intercept coincides with the origin.
2. Form the reciprocals of the above intercepts.
3. Express the reciprocal terms with the smallest set of integers that can be obtained by multiplying each of the fractions by the same number (i.e., reduce the terms to the smallest set of integer values possible).

Consider the (111) plane of Fig. 12.5. The plane intercepts the axes at unit axial lengths 1, 1, 1. The reciprocals of these unit values are naturally unity, and, since no common factors other than 1 are present, the Miller indexes are (111); hence the term (111)-oriented process.

The silicon-gate process is discussed in Section 12.3; Fig. 12.6 illustrates the sequence of the processing steps. In the (111) process, the equation that gives the threshold voltage is

$$V_{th} = \Phi_{MS} - \frac{Q_{ss}}{C_o} + 2\phi_f - \frac{Q_B}{C_o}$$

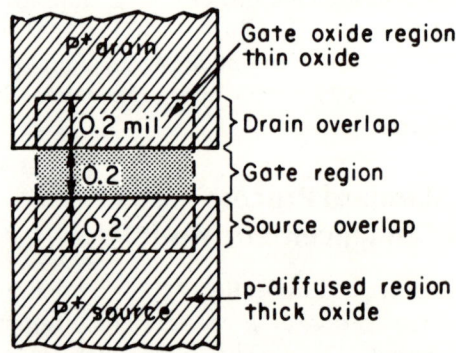

Figure 12.5  The silicon-gate process

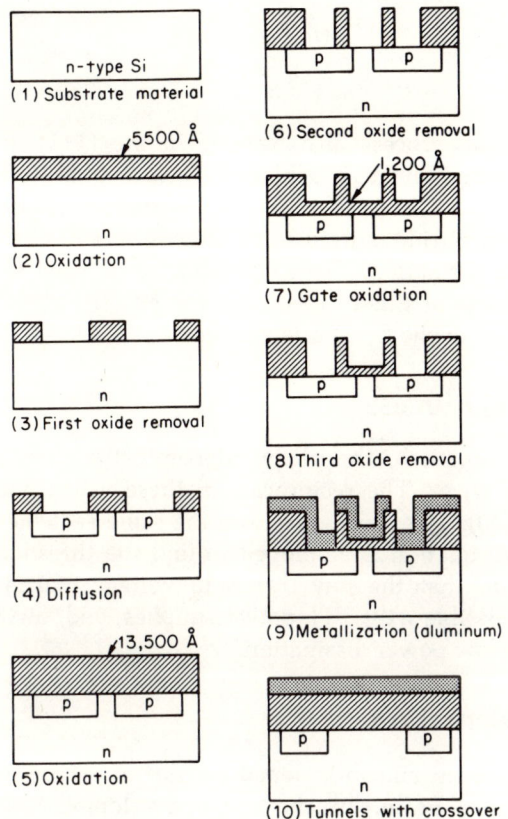

Figure 12.6 Standard process sequence

where

$\Phi_{MS}$ = work function between metal and silicon
$Q_{ss}$ = charge at the silicon–silicon dioxide interface
$C_o$ = capacitance of gate oxide
$\phi_f$ = Fermi potential

The power supply requirements for (111) standard process MOS/LSI are

$$V_{DD} = -12 \text{ V}$$
$$V_{GG} = -24 \text{ V}$$

The power supply requirements for the silicon-gate process are lower:

$$V_{DD} = -7.5 \text{ V}$$
$$V_{GG} = -15 \text{ V}$$

The nitride process also belongs in the (111) approach. It is employed to simultaneously achieve TTL compatibility and to exploit the high mobility obtained with (111) orientation. In this type of process, silicon nitride is used in conjunction with silicon dioxide for gate dielectric formation. The same equation is used to calculate the threshold voltage of this process, and the power supply requirements are the same as for the (111) orientation.

## 12.5 The (100) Process

In this approach, the plane intercepts the three axes at axial lengths of 1, ∞, ∞. The reciprocals of these values yield the Miller indexes of (100). The process follows the same sequence of Fig. 12.6, and the same equation is employed to find the threshold voltage. Its advantages are that the low threshold voltage makes this type of circuitry compatible with TTL power supplies, and, owing to this low supply, it has low power dissipation.

## 12.6 Ion Implantation

In this proces, silicon is doped with $n$- or $p$-type impurities by a high-energy ion beam. In the early development stages of this method, radiation damage resulting from the use of high-energy ion beams delayed the practical application of this technique. It was found that thermal annealing in conjunction with standard semiconductor processing techniques removed this unfavorable radiation. The two different methods of ion implantation useful in MOS/LSI process are (1) the self-aligned gate, and (2) channel doping. In both cases, boron ions are used for the implantation.

## 12.7 Memory Circuits

Large-scale integration has greatly contributed to the development of complex MOS and bipolar memory circuits. The reader will undoubtedly hear such terms as static and dynamic memory. A discussion of both types follows.

Basically, dynamic memories depend upon charge storage for data retention. If the clock stops, or runs too slow, in a dynamic

Memory Circuits    171

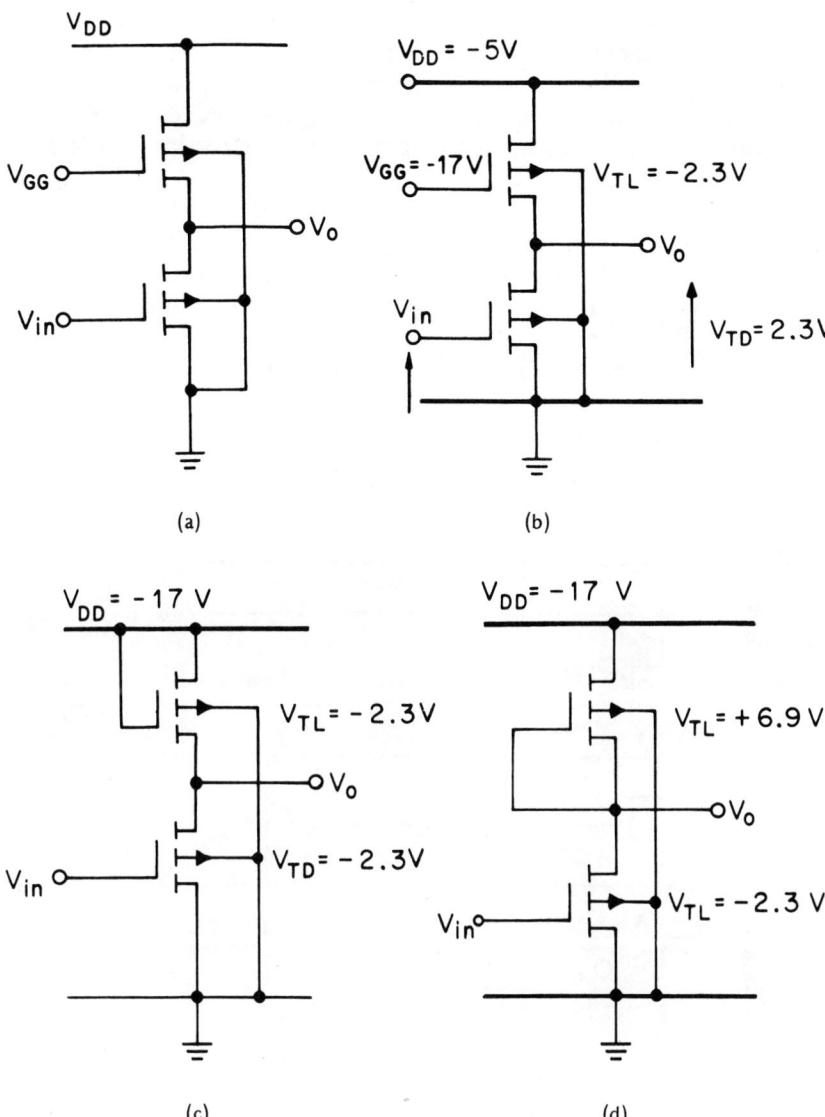

Figure 12.7    Four types of MOS circuits

memory, the information stored is lost. However, this deficiency is offset by the economy and speed of dynamic circuits.

Static memories, on the other hand, contain additional devices that form a bistable (static) latch within each delay element. The

## 172    Large-Scale Integration

static memory will not lose data when the clock is stopped. Static cells are generally larger and slower than dynamic cells, but do offer improved data-storage capabilities. Figure 12.7 illustrates four types of MOS circuits. All four cells are inverters. In part a, two MOS devices, $Q_1$ and $Q_2$, are connected as a static inverter. When input 1 is sufficiently high, $Q_2$ turns on and output 0 is low. If 1 is low, $Q_2$ is off and $Q_1$ pulls the output high. The circuit requires that $Q_1$ have drastically different geometry than $Q_2$. For equivalent bias, $Q_2$ must have much higher conductance than $Q_1$ to get reasonable noise margins. This is symbolized in both parts a and b by showing $Q_1$ as a resistor rather than as an active device. As a result of the low conductance of $Q_1$, current available from $Q_1$ for charging load capacitance is quite limited, and thus low-to-high transitions are rather slow.

Figure 12.7b is a variation of part a. When the clock pulse is active, the circuit behaves very much as in part a. By making the

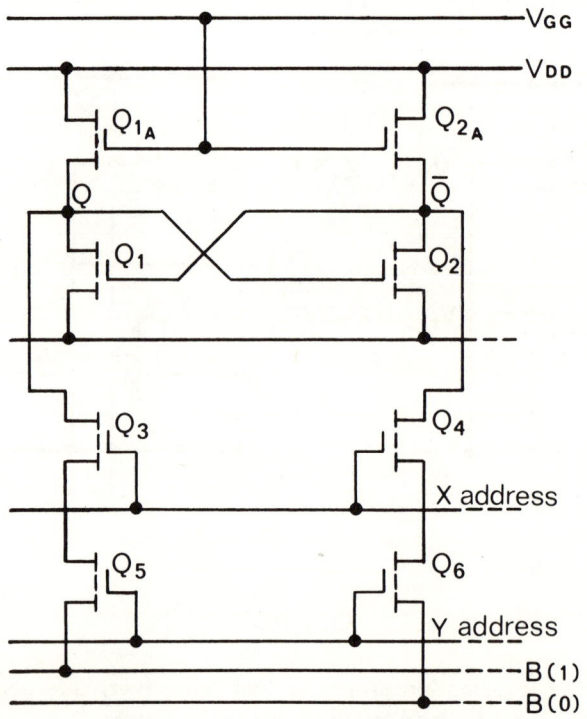

Figure 12.8    Static RAM cell

clock voltage higher than $V$, a more consistent high output level may be established. Once the output level is established, the clock may be turned off to save power.

In Fig. 12.7c, $Q_1$ has such a high inductance that, when the clock is active, low outputs may have noise margins so poor that they are unusable. However, after the clock is removed, low outputs quickly approach usable values if the input 1 is maintained long enough. During the time the clock is active, this circuit may consume large amounts of power (if input 1 is high).

In Fig. 12.7d, the capacitive load is charged when the clock is active (high), and is discharged when the clock returns to low, if and only if 1 is high. The circuit draws current only to charge and discharge the load capacitance; there is no dc power drain. However, the load capacitance is reflected back to the clock driver.

An example of a static random-access memory (RAM) cell is shown in Fig. 12.8. Two of the devices, $Q_{1A}$ and $Q_{2A}$, act as load resistors and are biased on by the $V_{GG}$ supply. $Q_1$ and $Q_2$ act as the bistable storage element; $Q_3$ through $Q_6$ are gating devices permitting one to address properly each separate cell within a large matrix circuit using a random-accessing scheme.

If we wish to read the contents of a cell in a particular $xy$ coordinate (address), we bring these two addressing lines into coinci-

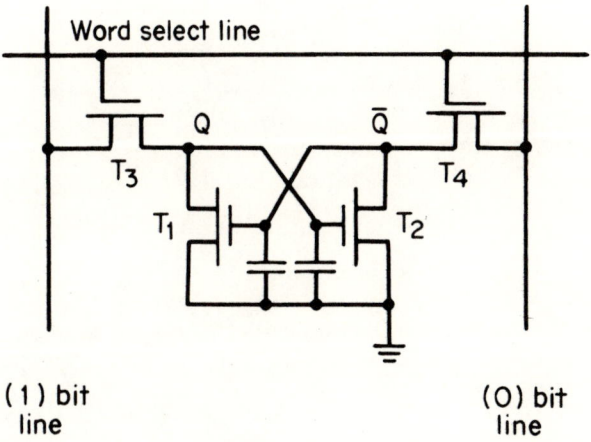

Figure 12.9  Dynamic RAM cell

dence with a negative-going voltage (the cell in discussion consists of $p$-channel enhancement-type devices). When the particular cell is addressed, $Q_3$ through $Q_6$ are turned on, permitting us to sense the relative voltages on two bit lines $B(0)$ and $B(1)$. These two bit lines are used to determine the state of each static cell. To write into the memory, we address the cell in the same manner, and set or reset the desired voltages onto two bit lines $B(0)$ and $B(1)$, which in turn forces the bistable storage element into the desired state.

A dynamic RAM cell is shown in Fig. 12.9. In contrast to the static memory cell, the word-enable transistors and the load devices, $Q_3$ and $Q_4$, are one and the same in this circuit. To read data from the cell, the word line is brought to a negative-voltage level, and the currents are sensed differentially between the two bit lines, which are kept in a negative-voltage state during the reading period. To write into the memory cell, the bit line is forced into the proper states as for the static cell, and then the data are transmitted into the cell by enabling the word line. To refresh the cell, the bit lines and the word lines are all brought to a negative potential.

## Exercises

**12.1** Describe some of the advantages of I²L.

**12.2** Describe the silicon-gate process.

**12.3** What is the difference between the (110) and the (111) orientation? Graph all three processes using a cube.

**12.4** What are the advantages and disadvantages of static and dynamic memory?

**12.5** Draw a memory cell and describe its operation.

**12.6** Given the following:

$$\Phi_{MS} = -0.3 \text{ V}$$
$$C_o = 2.9 \times 10^{-8} \text{ F/cm}^2$$
$$\phi_f = -0.29 \text{ V}$$
$$Q_{ss} = (5 \times 10^{11})(1.6 \times 10^{-19})$$

Find the threshold voltage of a device.

chapter thirteen
# Integrated-Circuit Applications

The previous chapters provide instruction on the actual design and fabrication of the integrated circuit. This chapter provides instruction on the use of IC's as building blocks of a system. After all, proper design of an IC greatly depends on its specific use after it has been fabricated.

## 13.1 The Operational Amplifier

One of the most commonly used four-terminal elements, and one of the most popular ones, is the operational amplifier. This element has applications in analog computers, logic circuits, feedback control systems, and many other electronic devices. In what follows, the basic operational amplifier is defined and some of its applications are discussed.

The symbolic representation of an op-amp is shown in Fig. 13.1a. Its corresponding idealized $v_2$–$v_1$ characteristic is shown in Fig. 13.1b. Note that for $v_1 \leqq E$, the output voltage $v_2$ is a linear function of $v_1$; that is, $v_2 = -Av_1$, where $A$ is the open-loop gain of the op-amp. For $v_1 \geqq E$, saturation occurs and the output remains constant. For actual op-amps, the gain ranges from 10,000 to 20,000. Thus

The Operational Amplifier     177

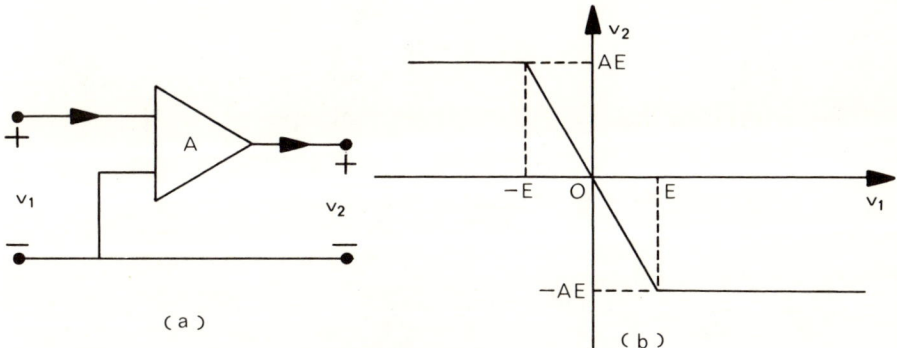

Figure 13.1  Op-amp symbol and characteristics.

small perturbations on the input voltage cause large changes in the output. For this reason, the use of op-amps in open-loop form is quite limited. Furthermore, since op-amps are built from transistors, the gain is quite sensitive to temperature and has a relatively large fluctuation.

The circuit realization of an open-loop op-amp is shown in Fig. 13.2a. Ideally, $R_1 \to \infty$ and $R_2 \to 0$. However, in practical op-amps, typical values are $R_1 = 50$ k$\Omega$ and $R_2 = 100$ $\Omega$. To remedy the shortcomings of the open-loop model, the output voltage is usually fed back through a resistor $R_f$ (Fig. 13.2b). To find the relationship between $v_1$ and $v_2$ for the feedback model, note that, since ideally the

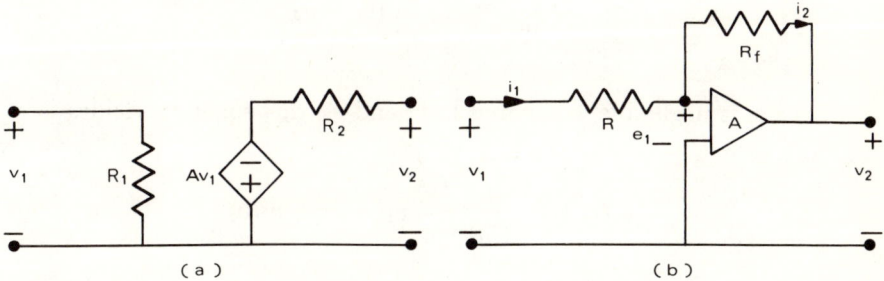

Figure 13.2  Open-loop op-amp.

input resistance of the open-loop op-amp is infinity, then, in Fig. 13.2b,

$$i_1 \simeq i_2$$

Hence it is clear that

$$\frac{v_1 - e_1}{R} = \frac{e_1 - v_2}{R_f}$$

However, $v_2 = -Ae_1$, and, consequently,

$$e_1 = -\frac{v_2}{A}$$

The gain of the feedback op-amp does not depend on the open-loop gain $A$. It only depends on the ratio $R_f/R$. This implies that the feedback gain is *insensitive* to the fluctuation of the open-loop gain. Some important applications of the feedback op-amp are given in the following examples.

**Example 1.** *Integrator*

Consider the feedback op-amp of Fig. 13.3a. The feedback consists of a capacitor, $C_f$. Since $i_1 \simeq i_2$, we can write

$$\frac{v_1 - e_1}{R} = C_f \frac{d}{dt}(e_1 - v_2)$$

Since $e_1$ is negligible, compared to $v_1$ and $v_2$, this equation yields

$$v_2(t) = -\frac{1}{C_f R}\int_0^t v_1(\tau)\,d\tau + v_2(0)$$

Assuming that $v_2(0) = 0$, this equation implies that the output voltage is proportional to the integral of the output voltage. Such integrators are extensively used in analog computers.

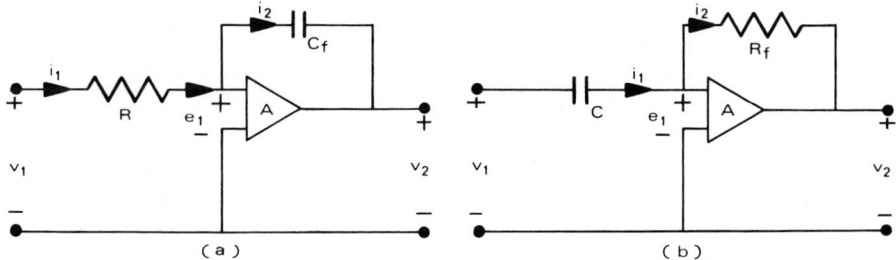

Figure 13.3

**Example 2.** *Differentiator*

Figure 13.3b shows an op-amp used as a differentiator. As in the previous example, since $i_1 \simeq i_2$ and $e_1$ is negligible compared to $v_1$ and $v_2$, we have

$$v_2(t) = -CR_f \frac{d}{dt} v_1(t)$$

That is, the output voltage is proportional to the derivative of the input voltage. Such differentiators, however, are not commonly used in practice owing to their susceptibility to random noise.

**Example 3.** *Converters*

Feedback op-amps can be used to convert a voltage source to a current source, and vice versa. The schematic representation of voltage-to-current and current-to-voltage converters is shown in Fig. 13.4a and b, respectively. In part a, since $i_1 \simeq i_2$, and $e_1$ is smaller than $v_1$, the current $i_2$ through the feedback can be written as

$$i_2 = \frac{1}{R_1} v_1$$

That is, $i_2$ is independent of $v_2$, and, therefore, the feedback path can be considered as an ideal current source whose current can be ad-

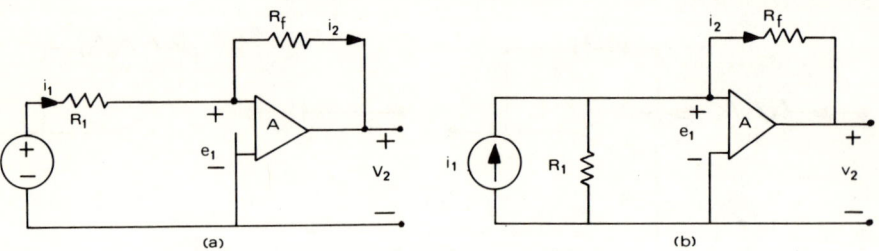

Figure 13.4  Converters.

justed by the voltage source $v_1$ and $R_1$. In Fig. 13.4b, the resistance of $R_1$ is chosen to be much smaller than $R_f$ so that $i_2 \simeq i_1$. Thus

$$i_1 \simeq i_2 = \frac{e_1 - v_2}{R_f} \simeq -\frac{1}{R_f} v_2$$

That is,

$$v_2 = -R_f i_1$$

Hence, the output of the amplifier can be considered as an ideal voltage source whose voltage can be adjusted by $R_f$ and the current source $i_1$. Naturally, these sources are ideal only on a limited range.

**Example 4.** *Adders*

Operational amplifiers can be used for adding several voltages (i.e., $v_1, v_2, \ldots, v_n$). This application of the op-amp is shown in Fig. 13.5a. Note that $i_1 \cong i_2 + \ldots + i_n$ or, equivalently,

$$\frac{e_1 - v_0}{R_f} \simeq \frac{v_1 - e_1}{R_1} + \frac{v_2 - e_1}{R_2} + \ldots + \frac{v_n - e_1}{R_n}$$

Since $e_1$ is negligible, we get

$$v_0 = -R_f \left( \frac{v_1}{R_1} + \frac{v_2}{R_2} + \ldots + \frac{v_n}{R_n} \right)$$

In particular,

if $R_1 = R_2 = \ldots = R_n = R_f$, then $v_0 = -(v_1 + v_2 + \ldots + v_n)$

**Example 5.** *Nonlinear Amplifiers*

If a nonlinear voltage-controlled resistor is placed in the forward path of an op-amp, the output voltage will be a nonlinear function of the input voltage. Consider the amplifier of Fig. 13.5b. Since $i_1 \simeq i_2$, we can write

$$i_1 = \frac{e_1 - v_2}{R_f} = f(v_1 - e_1)$$

Neglecting $e_1$, we get

$$v_2 = -f(v_1)$$

That is, the output is a nonlinear function of the input. As a numerical example, let $i_1 = -k(v_1 - e_1)^3$; then the $v_2 - v_1$ characteristic of the resulting op-amp becomes $v_2 = kv_1^3$.

Nonlinear amplifiers are usually used to compensate for the nonlinearities inherent in some existing elements. For instance, in a thermocouple sensing device, the voltage across the device may be given in terms of temperature as

$$v_1 = T^{1/8}$$

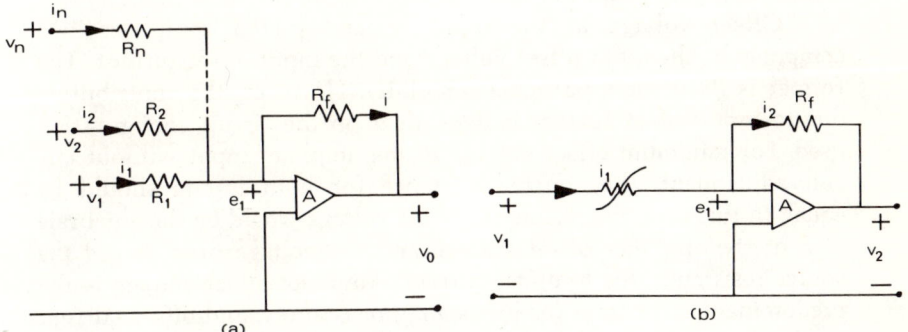

Figure 13.5  Non-linear amplifier.

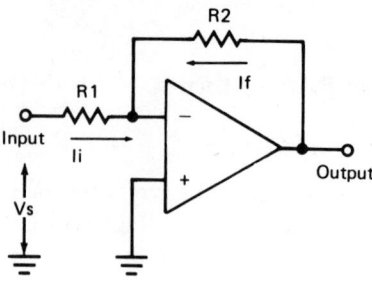

**Figure 13.6** Inverting amplifier.

Now, if the output of this device is connected in tandem to the nonlinear op-amp discussed previously, the output voltage $v_2$ will be a *linear* function of the temperature; that is, $v_2 = kv_1^3 = T$.

**Example 6.** *Inverting Amplifier*

The basic circuit is shown in Fig. 13.6. This circuit, as its name implies, inverts. It offers a closed-loop gain of $R_2/R_1$, provided that this ratio is small enough in comparison to the amplifier of open-loop gain. The input impedance is equal to $R_1$; the closed-loop bandwidth is equal to the unity-gain frequency divided by 1 plus the closed-loop gain.

The only precautions to keep in mind when using this circuit are (1) that $R_3$ should be chosen to equal the parallel combination of $R_1$ and $R_2$ to minimize the offset-voltage error due to bias current, and (2) there will be an offset voltage at the amplifier output equal to the closed-loop gain times the offset voltage at the amplifier input.

Offset voltage at the input of an op-amp comprises two components, the input offset voltage and the input offset current. The former is fixed for a particular amplifier. However, the contribution due to input offset current is dependent on the circuit configuration used. For minimum offset voltage at the amplifier input without circuit adjustment, the source resistance for both inputs should be equal. In this case, the maximum offset voltage would be the algebraic sum of the amplifier offset voltage and the voltage drop across the source resistance due to offset current. Amplifier offset voltage is the predominant error term for low-source resistances, and offset current produces the main error for high-source resistances.

In high-source-resistance applications, offset voltage at the amplifier output may be set by adjusting the value of $R_3$ and using the variation in voltage drop across it as an input-offset-voltage trim.

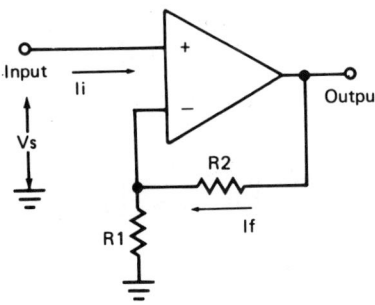

Figure 13.7 Non-inverting amplifier.

Offset voltage at the amplifier output is not as important in ac coupled applications. Here the only consideration is that any offset voltage at the output reduces the peak-to-peak linear output swing of the amplifier.

The gain-frequency characteristic of the amplifier and its feedback network can team up to cause oscillation. To avoid this condition, the phase shift through the amplifier and feedback network must never exceed 180° for any frequency where the gain of the amplifier and its feedback network are greater than unity. In practical applications, the phase shift should not approach 180°, since this is a situation of conditional stability. Obviously, the most critical case occurs when the attenuation of the feedback network is zero.

**Example 7.** *Noninverting Amplifier*

A high input impedance noninverting circuit is shown in Fig. 13.7. This circuit provides a closed-loop gain equal to the ratio of the sum of $R_1$ and $R_2$ to $R_1$ and a closed-loop 3-dB bandwidth equal to the amplifier unity-gain frequency divided by the closed-loop gain.

The primary differences between this circuit and the preceding one are (1) the output is not inverted, and (2) the input impedance is very high and is equal to the differential input impedance multiplied by the loop gain. In dc-coupled applications, input impedance is not as important as the input current and the voltage drop across the source resistance.

Application cautions are the same for this amplifier as for the inverting amplifier, with one exception. That is, the amplifier output will go into saturation if the input is allowed to float. This may be important if the amplifier must be switched from source to source. The compensation trade-off is stability versus bandwidth; larger val-

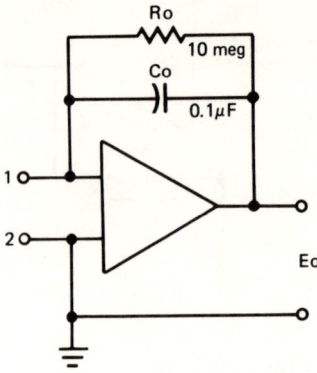

Figure 13.8 Unity-gain amplifier.

ues of compensation capacitor yield greater stability and lower bandwidth, and vice versa.

**Example 8.** *Unity-Gain Amplifier*

The unity-gain buffer is shown in Fig. 13.8. The circuit gives the highest input impedance of any op-amp circuit. Input impedance is equal to the differential input impedance multiplied by the open-loop gain, in parallel with common-mode input impedance. The gain error of this circuit is equal to the reciprocal of the amplifier open-loop gain or to the common-mode rejection, whichever is less.

Input impedance is a misleading concept in a dc-coupled unity-gain buffer. Bias current for the amplifier will be supplied by the source resistance and will cause an error at the amplifier input due to its voltage drop across the source resistance. Since this is the case, a low-bias-current amplifier should be chosen as a unity-gain buffer when working from high source resistances.

Three important considerations in the use of this circuit are (1) the amplifier must be compensated for unity-gain operation, (2) the output swing of the amplifier may be limited by the amplifier common-mode range, and (3) some amplifiers exhibit a latch-up mode when the amplifier common-mode range is exceeded.

**Example 9.** *Difference (Subtractor) Amplifier*

This type of circuit, shown in Fig. 13.9, is the complement of the adder. It allows the subtraction of two voltages or, as a special case, the cancellation of a signal common to the two inputs. The circuit is

The Operational Amplifier    185

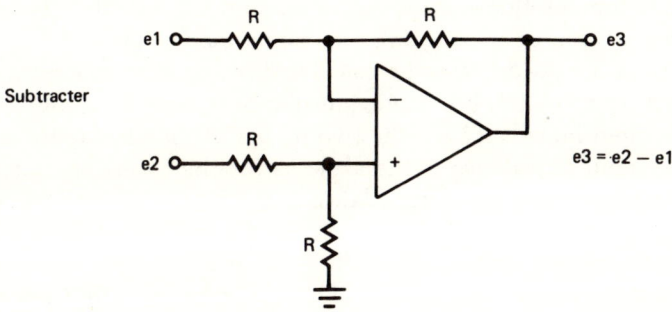

Figure 13.9   Subtractor.

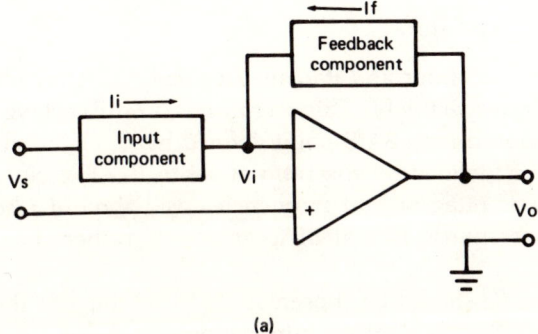

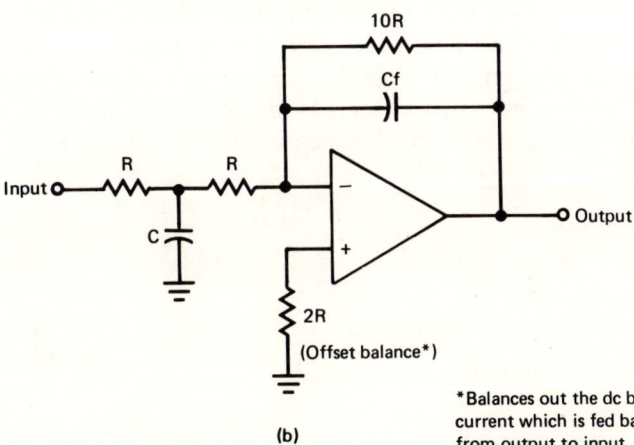

*Balances out the dc bias current which is fed back from output to input.

Figure 13.10   Low-pass filter.

useful as a computational amplifier, in making a differential to single-ended conversion, or in rejecting a common-mode signal.

Circuit bandwidth may be calculated in the same manner as for the inverting amplifier, but input impedance is somewhat more complicated. Input impedance for the two inputs is not necessarily equal; inverting input impedance is the same as for the inverting amplifier, and the noninverting input impedance is the sum of $R_3$ and $R_4$. Gain for either input is the ratio of $R_1$ to $R_2$ for the special case of a differential single-ended output, where $R_1 = R_3$ and $R_2 = R_4$. Compensation should be chosen on the basis of amplifier bandwidth. Care must be exercised in applying this circuit, since input impedances are not equal for minimum bias current error.

**Example 10.** *Low-Pass Filter*

A basic low-pass filter and its gain-frequency plot are shown in Fig. 13.10a and b, respectively. This circuit has a 6 dB/octave roll off after reaching a closed-loop 3-dB point defined by $f_c$. Gain below this corner frequency is defined by the ratio of $R_3$ to $R_1$. The circuit may be considered an ac integrator at frequencies well above $f_c$; however, the time-domain response is that of a single $RC$ rather than an integral.

The value of $R_2$ should be chosen so that it is equal to the parallel combination of $R_1$ and $R_3$, thus minimizing errors due to bias current. Either the amplifier should be compensated for unity gain or an internally compensated amplifier (the 741 type) should be used.

The preceding examples illustrate the versatility of the integrated op-amp, and provide a guide to a number of useful applications. The op-amp has been chosen here for discussion as a representative of linear integrated circuits precisely because of its wide applications.

# Exercises

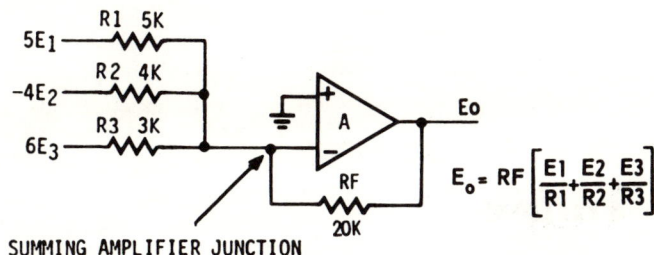

SUMMING AMPLIFIER JUNCTION

**13.1** The actual model of a feedback amplifier is shown (in the summing mode). Show that the voltage gain with feedback equals

$$A_f = \frac{R_f}{R_{in}} \times \frac{1}{1 + \frac{1}{A}\left(\frac{R_f}{R_{in}}\right)}$$

**13.2** For a typical op-amp, $A = 20{,}000$, $R_1 = 10$ K$\Omega$, $R_2 = 50$ k$\Omega$, $R_3 = 100$ $\Omega$, $R_f = 100$ k$\Omega$, and $R_L = 10$ k$\Omega$. For these values, show that $\varepsilon = 0.00065$.

**13.3** Find the output voltage of Figure 13.1 with resistor values as shown and with $E_1 = E_2 = E_3 = 5$ volts.

**13.4** Draw an inverting amplifier; designate the proper resistance values on your sketch.

**13.5** Repeat Exercise 13.4 for the noninverting amplifier.

# Index

Adders, 180
Advanced logic circuits, 143–145
Alpha cutoff frequency, 54
Aluminum:
  as conductor material, 18
  metallization, 81
Amplifiers:
  buffer, 67, 68
  Darlington amplifier, 115
  differential, 112–113
  and diodes, 45
  operational amplifiers, 113, 114, 176–185
  overall gain, 123
Analog computers, 176
Area, and wafer cost, 8
Artwork, fabrication, 95, 96
Attachment methods, 19

Ball bonding, 100
Bandwidth, 124
Base diffusion, 81
Base transport factor, calculated, 80

Biasing circuits, 117
Bipolar circuits, 127, 158, 162
  MOS devices, 149
  transistors, 46–48, 71
Bleeding capacity, and pinch resistors, 112
Bonding techniques, 90–100
  pads, 8
Boron diffusion, 25, 170
Breadboarding, 7, 95
Breakdown voltage, 79
  of pinch resistors, 111
Buffer amplifier, 67–68
Buried-layer technique, 53

Capacitance:
  calculation of, 85
  junction capacitance, 60, 78
  optimum, 86
  parasitic, 29, 38, 40, 83
  and pn junction, 4
  transition capacitance, 42, 80
  (*See also* Capacitors)

## Index

Capacitors, 1, 34–40
  clocking capacitors, 155
  design, 36, 84
  in integrated circuits, 104
  MOS capacitors, 39–40, 86
  nonpolarized, 38
  (*See also* Capacitance)
Carriers, 77, 92
Ceramic materials, 19
Channel doping method, 170
Channel inversion, 149
Charge recombination, 138
Chips, 37
  area, 155
  chip enable, 159
  size, 167
Circuit design (*see* Design)
Clock pulse, 172
Clocking capacitors, 155
Closed-loop biasing approach, 119
Collector-base diode, 42
Collector-base junction, 3, 34, 42, 81
Collector charging time, 64
Collector region, in transistors, 51
Collector resistance, calculated, 61, 63
Combination mode, of MOSFET, 150
Common base amplifiers, 113
Common drain circuit, 152
Common-gate configuration, 153
Common-mode rejection ratio, 113
Common source configuration, 152
Compensation, 45
Complementary metallic-oxide semiconductors (CMOS), 17, 148
Controlled-output current switch, 44
Converters, 179
Cost:
  of gates, 163
  of LSI circuits, 162
  of sapphire circuits, 17
  of wafer, 8
Crossovers, 8, 98, 115
Crystal structure, 14
Current calculation, graphical method, 121
Current capacity, of transistors, 54
Current flow, at junction, 47

Current gain:
  of input stage, 122
  in transistors, 86
Current-gain-bandwidth frequency, 54
Cut-and-peel plastic, 95
Cutoff frequency, 63
Czochralski method, 14

Darlington structure, 44, 115, 133, 136
Data storage, 158
  (*See also* Memory systems)
Depletion layer, 78
Depletion mode devices, 72, 150
Design, 53, 103–124
  complementary-type circuits, 16
  for monolithic circuits, 7
Difference amplifier, 185–186
Differential amplifier, 45, 112–113, 139
Differentiator, 179
Diffused resistors, 9, 24–25, 82
Diffused transistors, 48
Diffusion length, 77
Diodes:
  diode equation, 78
  fabrication, 3
  isolation effect, 98
  Schottky diodes, 109
  surface barrier diode, 138
Diode-resistor combination, 133
Diode-transistor logic (DTL), 44, 131
Direct-current bias currents, 120–121
Doping:
  with gold, 138
  impurities, 15
  with ion beam, 170
  level, 54, 79
  of substrate, 150
Dynamic logic, 158
Dynamic memories, 170–172

Economic factors, 162
Efficiency, of emitter, 79, 80
Emitter:
  crowding, 54, 92
  diffusion, 81, 96

## 190  Index

efficiency, 47, 79–80
followers, 140
stripes, 65–67
Emitter-base breakdown voltage, 108
Emitter-base junction, 34, 42
Emitter-coupled logic (ECL), 44, 137–142
ECL switch, 141, 142
Enhancement mode, of MOSFET, 150
Enhancement mode devices, 72, 151, 167
Epi layer, 52
Epitaxial capacitors, 35
Epitaxial diffused transistor, 10, 51
Epitaxial-layer resistors, 29
Epoxies, 18–19
Etching, and geometry, 30
Eutectic composition, 18
Exclusive-NOR function, 142

Fabrication, of integrated components, 81
Feedback control systems, 176
Feed op-amp, 178
Field-effect devices, 72
Films, 1, 11
Flip flops, 75, 143, 144
Float zone refining technique, 15
Forward-voltage equation, 79
Four-probe method, 15
Frequency, and resistor characteristics, 27
Frequency compensation, 104

Gain, of input stage, 122
Gallium, 17
Gaseous diffusion, 2
Gates, 127
 delay time, 131
 prices, 162
 transfer characteristics, 130
Geometry:
 and surface currents, 88
 of transistors, 86
Germanium, 13
Glass:
 as mask, 17
 as substrate material, 19

Gold, as conductor material, 18
Graphical calculation, of current, 121
Grounded configuration (*see* Common-gate configuration)
Growing technique (*see* Czochralski method)

High-current transistors, 92
Hole-electron pairs, recombination, 47

Impedance:
 calculation, 122
 and diffused resistor, 6
 output, 153
 of TTL circuits, 135
Impurity concentration, and resistance, 28–29
Input voltage, in TTL circuits, 136, 137
Integrated capacitors (*see* Capacitors)
Integrated diodes, 42–47
 collector-base diode, 42
Integrated-injection logic, 143, 163–164
Integrated-transistor structure, 51–53
Integrator, 178
Interconnect density, 165
Intraconnection patterns, 8
Intrinsic regions, 48
Inversion layer, p-type, 149
Inverter, 143, 164
 in MOS circuits, 155
Inverting amplifier, 183
Ion implantation, 170
Isolation, 37, 38
 capacitance, 60
 diffusion, 96, 155
 formation of, 81
 oxide isolation technique, 52, 53
 techniques, 6

J-K flip flop, 145
Junction capacitances, 60, 78
Junction capacitor, 39, 84–85

Junctions:
    equations, 2
    transistor action, 37

Kovar, 19–20

Large-scale integration (LSI), 162–174
Lateral pnp transistors, 105, 106
Lateral transistor effect, 10
Lifetime, of silicon, 77
Logic families, 127
Low-pass filter, 186
Low reverse voltage diode, 42
LSI (see Large-scale integration)

Mask:
    for amplifier circuit, 98–100
    fabrication, 95–100
Medium voltage diode, 42
Memory circuits, 71, 73, 75, 148, 170–172
Memory systems:
    and large-scale integration, 162
    and MOS devices, 158
Metal-gate MOS process, 167
Metal-insulator semiconductor (MIS), 148
Metal-oxide semiconductor forward effect transistor (MOSFET), 148
Metallic-oxide semiconductor (MOS) capacitors, 39–40
Metallic-oxide semiconductors, 71–75, 148
Metallization, 81, 96
Microphotography, 97
Microprocessors, 71, 148, 162, 167
Miller indexes, 168, 170
Monolithic fabrication, 4–7
    size, and yield, 9
MOS (see Metallic oxide semiconductor)
Mylar, 95

NAND gate, 127
NAND-NOR gate, 158
Negative logic, 127
Nitride process, 170

Noise immunity, 130–131
Noise-margin parameter, 131
Noninverting amplifier, 183
Nonlinear amplifiers, 181
Nonsaturated devices, 72
NOR gate, 127
NOT function, 156

Offset voltage, in op-amp, 183
Open-loop bandwidth, calculated, 123–124
Open-loop biasing approach, 118
Open-loop operational amplifier, 177
Operation enhancement mode devices, 72
Operational amplifier, 113, 114, 176
    schematic, 120
OR function, 141, 142
Output impedance, calculated, 153
Output structure, 135
Output voltage, in TTL circuits, 136, 137
Oxide capacitors, 9
Oxide isolation technique, 52, 53
Oxide thickness, 40

Packing materials, 20
Pads (see Bonding, pads)
Parallel arrangement, of MOS devices, 156
Parasitic capacitance (see Capacitance, parasitic)
Parasitic coupling, 6, 48–49
Parasitic resistance, in MOS devices, 149
Passive components, 153
Passive substrates, 1
Periphery-to-area ratio, 92
Phase splitter, 132
Photoengraving techniques, 54, 81, 95–100
Photolithographic techniques, 2
Photomasking, 166
Photoresist, 17
Pin connections, 98
Pinch-off voltage, 149
Pinch resistors, 109, 110
Polarization, of capacitors, 85
Positive logic, 127

# Index

Power consumption, in TTL circuits, 137
Power dissipation, 82
Precharge interval, 159
Preohmic areas, 96
Propagation delay time, 131

Radial gradient, 16
Radiation damage, 170
Random-access memory (RAM), 173
Random logic chips, 165
Recombination, of hole-electron pairs, 47
Rectifier, formation, 56–57
Reductions, 96–97
Refresh cycle, 160
Resistance:
   calculated, 24, 82, 83
   of emitter, 80
   end-to-end, 24
   four-probe measuring method, 16
   in monolithic circuits, 4
   and MOS transistors, 74
   of semiconductor materials, 15
Resistors:
   bleeding resistors, 141, 142
   collector FET resistor, 118
   design, 82
   epitaxial-layer (epi) resistor, 29
   pinch resistors, 109, 110
   silicon, 23
   thin-film resistors, 30, 83
   vapor-deposited resistor, 29–30
Resolution, of lens, 97
Reverse current equation, 79

Sapphire, 17
Saturation:
   collector-emitter, 10
   Harp resistor, 74
   of junction, 2
   in operational amplifier, 176
   of transistor, 149
Schottky-clamped structures, 109
Schottky outputs, 136
Schottky structure, 138, 163
   pure metal design, 164
Screen printing, 19
Self-aligned gate method, 170

Semiconductor monolithic circuits, 1
Semiconductors:
   CMOS circuits, 17
   indirect gap, 13
Series arrangement, of MOS logic, 156
Set-reset flip flop, 145
Sheet resistivity, 82
   of MOS process, 155
Silicon, 1, 2, 13–17, 77
   plane specification, 168
Silicon dioxide, 16–17
Silicon-gate process, MOS, 71–72, 166, 167
Source follower (*see* Common drain circuit)
Standby current, 115, 116
Static memories, 171–172
Step junction, 60
Substrated preparation, 81
Substrates:
   connection of, 99
   in MOS devices, 151
   silicon p type, 143
Subtraction amplifier (*see* Difference amplifier)
Superbeta transistors, 108, 109
Surface-effect devices, 71
Surface geometry, 88
   of transistor, 106
Surface recombination currents, 88
Switch (*see* Differential amplifier)
Switching rates, in ECL circuits, 138, 139
   for MOS devices, 151–152
   for operational amplifier, 176

Tantalum, 1, 18
Thermal compression bonding, 10
Thermal voltage, 2
Thin-film capacitors, 85
Thin-film process, 11
Thin-film resistors, 30, 82, 83
Three-electrode structure, 2
Threshold-level limits, 131
Threshold voltage (*see* Pinch-off voltage)
Topography, 7
Totem pole output structure, 135
Transition capacitance, 80

Transistor action, between junctions, 37
Transistor-transistor logic (TTL), 44, 73, 131–133
   output configurations, 133, 134
   unit loads (UL), 137
Transistors:
   beta, maximum, 59
   bipolar transistors, 46–48, 71
   design of, 86–92
   double base contact, 59
   emitter crowding, 92
   emitter efficiency, 79
   epitaxial diffused transistors, 51
   high current transistors, 65
   lateral pnp transistor, 104
   low power transistor, 65
   MOS transistors, 71
   multiple-emitter transistor, 131
   npn type, 2, 53
   npn-pnp combination, 56
   phase-splitter transistor, 132
   pnp transistors, 46, 49, 53
   p-n-p-n switches, 57, 58
   Schottky-clamped transistors, 109
   silicon-double-diffused planar-type, 2
   single-polarity type, 10
   size, 7
   superbeta transistor, 108, 109
Transconductance, 73
TTL (see Transistor-transistor logic)
Turn-off time, in Schottky structure, 109

Unipolar devices, 71
Unity-gain amplifier, 184

Vapor-deposited resistor, 29–30
Voltage follower circuit, 104

Wafers, 2
   intraconnection patterns, 8
Wire bonding, 18, 26, 99
Word-enable transistors, 174

Yield, 9

Zener diode, 44